Introduction to Bɪ

Introduction to Biomass

Edited by
Carter Campbell

Larsen & Keller
www.larsen-keller.com

Introduction to Biomass
Edited by Carter Campbell
ISBN: 978-1-63549-046-6 (Hardback)

© 2017 Larsen & Keller

 Larsen & Keller

Published by Larsen and Keller Education,
5 Penn Plaza,
19th Floor,
New York, NY 10001, USA

Cataloging-in-Publication Data

Introduction to biomass / edited by Carter Campbell.
    p. cm.
Includes bibliographical references and index.
ISBN 978-1-63549-046-6
1. Biomass. 2. Biomass energy. I. Campbell, Carter.
TP339 .I58 2017
662.88--dc23

The publisher's policy is to use permanent paper from mills that operate a sustainable forestry policy. Furthermore, the publisher ensures that the text paper and cover boards used have met acceptable environmental accreditation standards.

Printed and bound in the United States of America.

For more information regarding Larsen and Keller Education and its products, please visit the publisher's website www.larsen-keller.com

Table of Contents

Preface

This book is a compilation of chapters that discuss the most vital concepts in the field of biomass. It aims to provide the readers with thorough knowledge about this subject. Biomass is organic matter which is produced from living organisms, mainly from plants. It is used in two ways, either directly or by converting it into biofuel. This text is a valuable compilation of topics, ranging from the basic to the most complex theories and principles in the area of biomass. Through this book, we attempt to further enlighten the readers about the various techniques and technologies associated with this field. Different approaches, evaluations and methodologies have been included in it. This textbook is an essential guide for graduates and post-graduates in this field.

A short introduction to every chapter is written below to provide an overview of the content of the book:

Chapter 1 - Biomass is the fuel that is derived from living organisms. Biomass is either used directly or indirectly, directly it can be used by combustion to produce heat whereas indirectly it is converted to various forms. This chapter is an overview of the subject matter incorporating all the major aspects of biomass.; **Chapter 2** - Bioenergy is a renewable source of energy; it stores sunlight in the form of chemical energy. Bioenergy comprises wood, wood waste, manure and other products from agricultural processes. Bioconversion, biomass heating system and biofuels are some aspects of bioenergy that have been discussed in the following section; **Chapter 3** - The applications of biomass dealt within this chapter are biogas, biochar, bioliquids, blue carbon, straw and woodchips. The mixture of gases produced by the breaking of organic matter is known as biogas. Biochar is used as a product that is added to soil to improve it. The various aspects elucidated in this text are of vital importance, and provides a better understanding of biomass ; **Chapter 4** - Plants in a number of ways are utilized in biomass. Some of the applications are lignocellulosic biomass, energy crop and panicum virgatum. Plant dry matter is referred to as lignocellulosic biomass whereas plants with little maintenance are used as biofuels are called energy crops. This section will provide an integrated understanding on the utilization of plants in biomass production;

Finally, I would like to thank my fellow scholars who gave constructive feedback and my family members who supported me at every step.

Editor

Introduction to Biomass

Biomass is the fuel that is derived from living organisms. Biomass is either used directly or indirectly, directly it can be used by combustion to produce heat whereas indirectly it is converted to various forms. This chapter is an overview of the subject matter incorporating all the major aspects of biomass.

Biomass

Sugarcane plantation in Brazil. Sugarcane bagasse is a type of biomass.

Biomass is organic matter derived from living, or recently living organisms. Biomass can be used as a source of energy and it most often refers to plants or plant-based materials which are not used for food or feed, and are specifically called lignocellulosic biomass. As an energy source, biomass can either be used directly via combustion to produce heat, or indirectly after converting it to various forms of biofuel. Conversion of biomass to biofuel can be achieved by different methods which are broadly classified into: *thermal*, *chemical*, and *biochemical* methods.

Biomass Sources

Historically, humans have harnessed biomass-derived energy since the time when people began burning wood to make fire. Even today, biomass is the only source of fuel for domestic use in many developing countries. Biomass is all biologically-produced matter based in carbon, hydrogen and oxygen. The estimated biomass production in the world is 104.9 petagrams (104.9 * 10^{15} g - about 105 billion metric tons) of carbon per year, about half in the ocean and half on land.

Wood remains the largest biomass energy source today; examples include forest residues (such as dead trees, branches and tree stumps), yard clippings, wood chips and even municipal solid waste. Wood energy is derived by using lignocellulosic biomass (second-generation biofuels) as fuel. Harvested wood may be used directly as a fuel or collected from wood waste streams. The largest source of energy from wood is pulping liquor or "black liquor," a waste product from processes of the pulp, paper and paperboard industry. In the second sense, biomass includes plant or animal matter that can be converted into fibers or other industrial chemicals, including biofuels. Industrial biomass can be grown from numerous types of plants, including miscanthus, switchgrass, hemp, corn, poplar, willow, sorghum, sugarcane, bamboo, and a variety of tree species, ranging from eucalyptus to oil palm (palm oil).

Eucalyptus in Brazil. Remains of the tree are reused for power generation.

Based on the source of biomass, biofuels are classified broadly into two major categories. First-generation biofuels are derived from sources such as sugarcane and corn starch. Sugars present in this biomass are fermented to produce bioethanol, an alcohol fuel which can be used directly in a fuel cell to produce electricity or serve as an additive to gasoline. However, utilizing food-based resources for fuel production only aggravates the food shortage problem. Second-generation biofuels, on the other hand, utilize non-food-based biomass sources such as agriculture and municipal waste. These biofuels mostly consist of lignocellulosic biomass, which is not edible and is a low-value waste for many industries. Despite being the favored alternative, economical production of second-generation biofuel is not yet achieved due to technological issues. These issues arise mainly due to chemical inertness and structural rigidity of lignocellulosic biomass.

Plant energy is produced by crops specifically grown for use as fuel that offer high biomass output per hectare with low input energy. Some examples of these plants are wheat, which typically yields 7.5–8 tonnes of grain per hectare, and straw, which typically yields 3.5–5 tonnes per hectare in the UK. The grain can be used for liquid transportation fuels while the straw can be burned to produce heat or electricity. Plant biomass can also be degraded from cellulose to glucose through a series of chemical treatments, and the resulting sugar can then be used as a first-generation biofuel.

The main contributors of waste energy are municipal solid waste, manufacturing waste, and landfill gas. Energy derived from biomass is projected to be the largest non-hydro-electric renewable resource of electricity in the US between 2000 and 2020.

Biomass can be converted to other usable forms of energy like methane gas or transportation fuels like ethanol and biodiesel. Rotting garbage, and agricultural and human waste, all release methane gas, also called landfill gas or biogas. Crops such as corn and sugarcane can be fermented to produce the transportation fuel ethanol. Biodiesel, another transportation fuel, can be produced from leftover food products like vegetable oils and animal fats. Also, biomass-to-liquids (called "BTLs") and cellulosic ethanol are still under research.

There is research involving algae, or algae-derived, biomass, as this non-food resource can be produced at rates five to ten times those of other types of land-based agriculture, such as corn and soy. Once harvested, it can be fermented to produce biofuels such as ethanol, butanol, and methane, as well as biodiesel and hydrogen. Efforts are being made to identify which species of algae are most suitable for energy production. Genetic engineering approaches could also be utilized to improve microalgae as a source of biofuel.

The biomass used for electricity generation varies by region. Forest by-products, such as wood residues, are common in the US. Agricultural waste is common in Mauritius (sugar cane residue) and Southeast Asia (rice husks). Animal husbandry residues, such as poultry litter, are common in the UK.

As of 2015, a new bioenergy sewage treatment process aimed at developing countries is under trial; the Omni Processor is a self-sustaining process which uses sewerage solids as fuel in a process to convert waste water into drinking water, with surplus electrical energy being generated for export.

Comparison of Total Plant Biomass Yields (Dry Basis)

World Resources

If the total annual primary production of biomass is just over 100 billion (1.0E+11) tonnes of Carbon /yr, and the energy reserve per metric tonne of biomass is between about 1.5E3 – 3E3 Kilowatt hours (5E6 – 10E6 BTU), or 24.8 TW average, then biomass could in principle provide 1.4 times the approximate annual 150E3 Terrawatt.hours required for the current world energy consumption. For reference, the total solar power on Earth is 174 kTW. The biomass equivalent to solar energy ratio is 143 ppm (parts per million), given current living system coverage on Earth. Best in class solar cell efficiency is (20-40)%. Additionally, Earth's internal radioactive energy production, largely the driver for volcanic activity, continental drift, etc., is in the same range of power, 20 TW. At some 50% carbon mass content in biomass, annual production, this corresponds to about 6% of atmospheric carbon content in CO_2 (for the current 400 ppm).

(1.0E+11 tonnes biomass annually produced approximately 25 TW)

Annual world biomass energy equivalent =16.7 - 33.4 TW.

Annual world energy consumption =17.7. On average, biomass production is 1.4 times larger than world energy consumption.

Common Commodity Food Crops

- Agave: 1–21 tons/acre
- Alfalfa: 4–6 tons/acre
- Barley: grains – 1.6–2.8 tons/acre, straw – 0.9–2.5 tons/acre, total – 2.5–5.3 tons/acre
- Corn: grains – 3.2–4.9 tons/acre, stalks and stovers – 2.3–3.4 tons/acre, total – 5.5–8.3 tons/acre
- Jerusalem artichokes: tubers 1–8 tons/acre, tops 2–13 tons/acre, total 9–13 tons/acre
- Oats: grains – 1.4–5.4 tons/acre, straw – 1.9–3.2 tons/acre, total – 3.3–8.6 tons/acre
- Rye: grains – 2.1–2.4 tons/acre, straw – 2.4–3.4 tons/acre, total – 4.5–5.8 tons/acre
- Wheat: grains – 1.2–4.1 tons/acre, straw – 1.6–3.8 tons/acre, total – 2.8–7.9 tons/acre

Woody Crops

- Oil palm: fronds 11 ton/acre, whole fruit bunches 1 ton/acre, trunks 30 ton/acre

Not Yet in Commercial Planting

- Giant miscanthus: 5–15 tons/acre
- Sunn hemp: 4.5 tons/acre
- Switchgrass: 4–6 tons/acre

Genetically Modified Varieties

- Energy Sorghum

Biomass Conversion

Thermal Conversion

Thermal conversion processes use heat as the dominant mechanism to convert biomass into another chemical form. The basic alternatives of combustion (torrefaction, pyrolysis, and gasification) are separated principally by the extent to which the chemical reactions involved are allowed to proceed (mainly controlled by the availability of oxygen and conversion temperature).

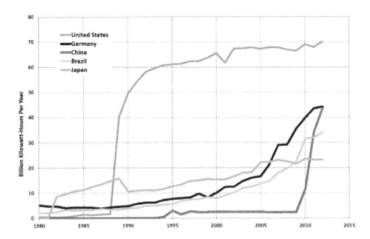

Trends in the top five countries generating electricity from biomass

Energy created by burning biomass (fuel wood) is particularly suited for countries where the fuel wood grows more rapidly, e.g. tropical countries. There are a number of other less common, more experimental or proprietary thermal processes that may offer benefits such as hydrothermal upgrading (HTU) and hydroprocessing. Some have been developed for use on high moisture content biomass, including aqueous slurries, and allow them to be converted into more convenient forms. Some of the applications of thermal conversion are combined heat and power (CHP) and co-firing. In a typical dedicated biomass power plant, efficiencies range from 20–27% (higher heating value basis). Biomass cofiring with coal, by contrast, typically occurs at efficiencies near those of the coal combustor (30–40%, higher heating value basis).

Biomass briquettes are an example fuel for production of dendrothermal energy

Chemical Conversion

A range of chemical processes may be used to convert biomass into other forms, such as to produce a fuel that is more conveniently used, transported or stored, or to exploit some property of the process itself. Many of these processes are based in large part on similar coal-based processes, such as Fischer-Tropsch synthesis, methanol production, olefins (ethylene and propylene), and similar chemical or fuel feedstocks. In most cases, the first step involves gasification, which step generally is the most expensive and involves the greatest technical risk. Biomass is more difficult to feed into a pressure vessel than coal or any liquid. Therefore, biomass gasification is frequently done at atmospheric pressure and causes combustion of biomass to produce a combustible gas consisting of carbon monoxide, hydrogen, and traces of methane. This gas mixture, called a producer gas, can provide fuel for various vital processes, such as internal combustion engines, as well as substitute for furnace oil in direct heat applications. Because any biomass material can undergo gasification, this process is far more attractive than ethanol or biomass production, where only particular biomass materials can be used to produce a fuel. In addition, biomass gasification is a desirable process due to the ease at which it can convert solid waste (such as wastes available on a farm) into producer gas, which is a very usable fuel.

Conversion of biomass to biofuel can also be achieved via selective conversion of individual components of biomass. For example, cellulose can be converted to intermediate platform chemical such a sorbitol, glucose, hydroxymethylfurfural etc. These chemical are then further reacted to produce hydrogen or hydrocarbon fuels.

Biomass also has the potential to be converted to multiple commodity chemicals. Halomethanes have successfully been by produced using a combination of A. fermentans and engineered S. cerevisiae. This method converts NaX salts and unprocessed biomass such as switchgrass, sugarcane, corn stover, or poplar into halomethanes. S-adenosylmethionine which is naturally occurring in S. cerevisiae allows a methyl group to be transferred. Production levels of 150 mg $L-1H-1$ iodomethane were achieved. At these levels roughly 173000L of capacity would need to be operated just to replace the United States' need for iodomethane. However, an advantage of this method is that it uses NaI rather than I2; NaI is significantly less hazardous than I2. This method may be applied to produce ethylene in the future.

Other chemical processes such as converting straight and waste vegetable oils into biodiesel is transesterification.

Biochemical Conversion

As biomass is a natural material, many highly efficient biochemical processes have developed in nature to break down the molecules of which biomass is composed, and many of these biochemical conversion processes can be harnessed.

Biochemical conversion makes use of the enzymes of bacteria and other microorganisms to break down biomass. In most cases, microorganisms are used to perform the conversion process: anaerobic digestion, fermentation, and composting.

Electrochemical Conversion

In addition to combustion, bio-mass/bio-fuels can be directly converted to electrical energy via electrochemical oxidation of the material. This can be performed directly in a direct carbon fuel cell, direct ethanol fuel cell or a microbial fuel cell. The fuel can also be consumed indirectly via a fuel cell system containing a reformer which converts the bio-mass into a mixture of CO and H2 before it is consumed in the fuel cell.

In the United States

The biomass power generating industry in the United States, which consists of approximately 11,000 MW of summer operating capacity actively supplying power to the grid, produces about 1.4 percent of the U.S. electricity supply.

Currently, the New Hope Power Partnership is the largest biomass power plant in the U.S. The 140 MW facility uses sugarcane fiber (bagasse) and recycled urban wood as fuel to generate enough power for its large milling and refining operations as well as to supply electricity for nearly 60,000 homes.

Second-generation Biofuels

Second-generation biofuels were not (in 2010) produced commercially, but a considerable number of research activities were taking place mainly in North America, Europe and also in some emerging countries. These tend to use feedstock produced by rapidly reproducing enzymes or bacteria from various sources including excrement grown in Cell cultures or hydroponics There is huge potential for second generation biofuels but non-edible feedstock resources are highly under-utilized.

Environmental Impact

Using biomass as a fuel produces air pollution in the form of carbon monoxide, carbon dioxide, NOx (nitrogen oxides), VOCs (volatile organic compounds), particulates and other pollutants at levels above those from traditional fuel sources such as coal or natural gas in some cases (such as with indoor heating and cooking). Utilization of wood biomass as a fuel can also produce fewer particulate and other pollutants than open burning as seen in wildfires or direct heat applications. Black carbon – a pollutant created by combustion of fossil fuels, biofuels, and biomass – is possibly the second largest contributor to global warming. In 2009 a Swedish study of the giant brown haze that periodically covers large areas in South Asia determined that it had been principally produced by biomass burning, and to a lesser extent by fossil-fuel burning. Researchers measured a significant concentration of C, which is associated with recent plant life rather than with fossil fuels.

Biomass power plant size is often driven by biomass availability in close proximity as transport costs of the (bulky) fuel play a key factor in the plant's economics. It has to be noted, however, that rail and especially shipping on waterways can reduce transport costs significantly, which has led to a global biomass market. To make small plants of 1 MWel economically profitable those power plants need to be equipped with technology that is able to convert biomass to useful electricity with high efficiency such as ORC technology, a cycle similar to the water steam power process just with an organic working medium. Such small power plants can be found in Europe.

On combustion, the carbon from biomass is released into the atmosphere as carbon dioxide (CO_2). The amount of carbon stored in dry wood is approximately 50% by weight. However, according to the Food and Agriculture Organization of the United Nations, plant matter used as a fuel can be replaced by planting for new growth. When the biomass is from forests, the time to recapture the carbon stored is generally longer, and the carbon storage capacity of the forest may be reduced overall if destructive forestry techniques are employed.

Industry professionals claim that a range of issues can affect a plant's ability to comply with emissions standards. Some of these challenges, unique to biomass plants, include inconsistent fuel supplies and age. The type and amount of the fuel supply are completely reliant factors; the fuel can be in the form of building debris or agricultural waste (such as deforestation of invasive species or orchard trimmings). Furthermore, many of the biomass plants are old, use outdated technology and were not built to comply with today's stringent standards. In fact, many are based on technologies developed during the term of U.S. President Jimmy Carter, who created the United States Department of Energy in 1977.

The U.S. Energy Information Administration projected that by 2017, biomass is expected to be about twice as expensive as natural gas, slightly more expensive than nuclear power, and much less expensive than solar panels. In another EIA study released, concerning the government's plan to implement a 25% renewable energy standard by 2025, the agency assumed that 598 million tons of biomass would be available, accounting for 12% of the renewable energy in the plan.

The adoption of biomass-based energy plants has been a slow but steady process. Between the years of 2002 and 2012 the production of these plants has increased 14%. In the United States, alternative electricity-production sources on the whole generate about 13% of power; of this fraction, biomass contributes approximately 11% of the alternative production. According to a study conducted in early 2012, of the 107 operating biomass plants in the United States, 85 have been cited by federal or state regulators for the violation of clean air or water standards laws over the past 5 years. This data also includes minor infractions.

Despite harvesting, biomass crops may sequester carbon. For example, soil organic carbon has been observed to be greater in switchgrass stands than in cultivated crop-land soil, especially at depths below 12 inches. The grass sequesters the carbon in its increased root biomass. Typically, perennial crops sequester much more carbon than annual crops due to much greater non-harvested living biomass, both living and dead, built up over years, and much less soil disruption in cultivation.

The proposal that biomass is carbon-neutral put forward in the early 1990s has been superseded by more recent science that recognizes that mature, intact forests sequester carbon more effectively than cut-over areas. When a tree's carbon is released into the atmosphere in a single pulse, it contributes to climate change much more than wood-land timber rotting slowly over decades. Current studies indicate that "even after 50 years the forest has not recovered to its initial carbon storage" and "the optimal strategy is likely to be protection of the standing forest".

The pros and cons of biomass usage regarding carbon emissions may be quantified with the ILUC factor. There is controversy surrounding the usage of the ILUC factor.

Forest-based biomass has recently come under fire from a number of environmental organizations, including Greenpeace and the Natural Resources Defense Council, for the harmful impacts it can have on forests and the climate. Greenpeace recently released a report entitled "Fuelling a BioMess" which outlines their concerns around forest-based biomass. Because any part of the tree can be burned, the harvesting of trees for energy production encourages Whole-Tree Harvesting, which removes more nutrients and soil cover than regular harvesting, and can be harmful to the long-term health of the forest. In some jurisdictions, forest biomass removal is increasingly involving elements essential to functioning forest ecosystems, including standing trees, naturally disturbed forests and remains of traditional logging operations that were previously left in the forest. Environmental groups also cite recent scientific research which has found that it can take many decades for the carbon released by burning biomass to be recaptured by regrowing trees, and even longer in low productivity areas; furthermore, logging operations may disturb forest soils and cause them to release stored carbon. In light of the pressing need to reduce greenhouse gas emissions in the short term in order to mitigate the effects of climate change, a number of environmental groups are opposing the large-scale use of forest biomass in energy production.

Supply Chain Issues

With the seasonality of biomass supply and a great variability in sources, supply chains play a key role in cost-effective delivery of bioenergy. There are several potential challenges peculiar to bioenergy supply chains:

Technical issues

- Inefficiencies of conversion

- Storage methods for seasonal availability

- Complex multi-component constituents incompatible with maximizing efficiency of single purpose use

- High water content

- Conflicting decisions (technologies, locations, and routes)

- Complex location analysis (source points, inventory facilities, and production plants)

Logistic issues

- Seasonal availability leading to storage challenges and/or seasonally idle facilities

- Low bulk-density and/or high water content

- Finite productivity per area and/or time incompatible with conventional approach to economy of scale focusing on maximizing facility size

Financial issues

- The limits for the traditional approach to economy of scale which focuses on maximizing single facility size

- Unavailability and complexity of life cycle costing data

- Lack of required transport infrastructure

- Limited flexibility or inflexibility to energy demand

- Risks associated with new technologies (insurability, performance, rate of return)

- Extended market volatilities (conflicts with alternative markets for biomass)

- Difficult or impossible to use financial hedging methods to control cost

Social issues

- Lack of participatory decision making

- Lack of public/community awareness

- Local supply chain impacts vs. global benefits

- Health and safety risks

- Extra pressure on transport sector

- Decreasing the esthetics of rural areas

Policy and regulatory issues

- Impact of fossil fuel tax on biomass transport

- Lack of incentives to create competition among bioenergy producers

- Focus on technology options and less attention to selection of biomass materials

- Lack of support for sustainable supply chain solutions

Institutional and organizational issues

- Varied ownership arrangements and priorities among supply chain parties

- Lack of supply chain standards

- Impact of organizational norms and rules on decision making and supply chain coordination

- Immaturity of change management practices in biomass supply chains

Biomass (Ecology)

Apart from bacteria, the total global live biomass has been estimated as 560 billion tonnes C, most of which is found in forests.

Biomass, in ecology, is the mass of living biological organisms in a given area or ecosystem at a given time. Biomass can refer to *species biomass*, which is the mass of one or more species, or to *community biomass*, which is the mass of all species in the community. It can include microorganisms, plants or animals. The mass can be expressed as the average mass per unit area, or as the total mass in the community.

Shallow aquatic environments, such as wetlands, estuaries and coral reefs, can be as productive as forests, generating similar amounts of new biomass each year on a given area.

How biomass is measured depends on why it is being measured. Sometimes, the biomass is regarded as the natural mass of organisms *in situ*, just as they are. For example, in a salmon fishery, the salmon biomass might be regarded as the total wet weight the salmon would have if they were taken out of the water. In other contexts, biomass can be measured in terms of the dried organic mass, so perhaps only 30% of the actual weight might count, the rest being water. For other purposes, only biological tissues count, and teeth, bones and shells are excluded. In some applications, biomass is measured as the mass of organically bound carbon (C) that is present.

Apart from bacteria, the total live biomass on Earth is about 560 billion tonnes C, and the total annual primary production of biomass is just over 100 billion tonnes C/yr. The total live biomass of bacteria may be as much as that of plants and animals or may be much less. The total amount of DNA base pairs on Earth, as a possible approximation of global biodiversity, is estimated at 5.0×10^{37}, and weighs 50 billion tonnes. In comparison, the total mass of the biosphere has been estimated to be as much as 4 TtC (trillion tons of carbon).

Ecological Pyramids

An ecological pyramid is a graphical representation that shows, for a given ecosystem, the relationship between biomass or biological productivity and trophic levels.

- A *biomass pyramid* shows the amount of biomass at each trophic level.

- A *productivity pyramid* shows the production or turn-over in biomass at each trophic level.

An ecological pyramid provides a snapshot in time of an ecological community.

The bottom of the pyramid represents the primary producers (autotrophs). The primary producers take energy from the environment in the form of sunlight or inorganic chemicals and use it to create energy-rich molecules such as carbohydrates. This mechanism is called primary production. The pyramid then proceeds through the various trophic levels to the apex predators at the top.

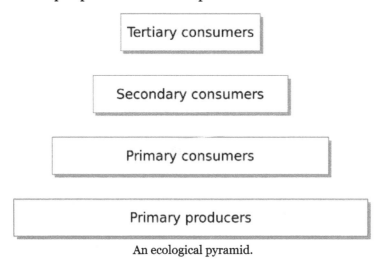
An ecological pyramid.

When energy is transferred from one trophic level to the next, typically only ten percent is used to build new biomass. The remaining ninety percent goes to metabolic processes or is dissipated as heat. This energy loss means that productivity pyramids are never inverted, and generally limits food chains to about six levels. However, in oceans, biomass pyramids can be wholly or partially inverted, with more biomass at higher levels.

Terrestrial Biomass

Terrestrial biomass generally decreases markedly at each higher trophic level (plants, herbivores, carnivores). Examples of terrestrial producers are grasses, trees and shrubs. These have a much higher biomass than the animals that consume them, such as deer, zebras and insects. The level with the least biomass are the highest predators in the food chain, such as foxes and eagles.

In a temperate grassland, grasses and other plants are the primary producers at the bottom of the pyramid. Then come the primary consumers, such as grasshoppers, voles and bison, followed by the secondary consumers, shrews, hawks and small cats. Finally the tertiary consumers, large cats and wolves. The biomass pyramid decreases markedly at each higher level.

Ocean Biomass

The Marine Food Chain

Phytoplankton are the main primary producers at the bottom of the marine food chain. Phytoplankton use photosynthesis to convert inorganic carbon into protoplasm. They are then consumed by microscopic animals called zooplankton.

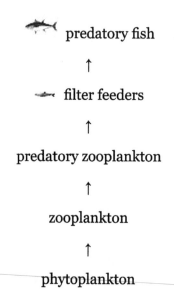

predatory fish

↑

filter feeders

↑

predatory zooplankton

↑

zooplankton

↑

phytoplankton

Ocean biomass, in a reversal of terrestrial biomass, can increase at higher trophic levels. In the ocean, the food chain typically starts with phytoplankton, and follows the course:

Phytoplankton → zooplankton → predatory zooplankton → filter feeders → predatory fish

Zooplankton comprise the second level in the food chain, and includes small crustaceans, such as copepods and krill, and the larva of fish, squid, lobsters and crabs.

In turn, small zooplankton are consumed by both larger predatory zooplankters, such as krill, and by forage fish, which are small schooling filter feeding fish. This makes up the third level in the food chain.

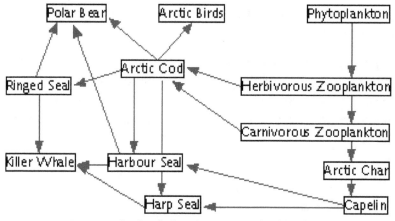

An ocean food web showing a network of food chains

The fourth trophic level consists of predatory fish, marine mammals and seabirds that consume forage fish. Examples are swordfish, seals and gannets.

Apex predators, such as orcas, which can consume seals, and shortfin mako sharks, which can consume swordfish, make up the fifth trophic level. Baleen whales can consume zooplankton and krill directly, leading to a food chain with only three or four trophic levels.

Marine environments can have inverted biomass pyramids. In particular, the biomass of consumers (copepods, krill, shrimp, forage fish) is larger than the biomass of primary producers. This happens because the ocean's primary producers are tiny phytoplankton that grow and reproduce rapidly, so a small mass can have a fast rate of primary production. In contrast, terrestrial primary producers grow and reproduce slowly.

There is an exception with cyanobacteria. Marine cyanobacteria are the smallest known photosynthetic organisms; the smallest of all, *Prochlorococcus*, is just 0.5 to 0.8 micrometres across. Prochlorococcus is possibly the most plentiful species on Earth: a single millilitre of surface seawater may contain 100,000 cells or more. Worldwide, there are estimated to be several octillion ($\sim 10^{27}$) individuals. *Prochlorococcus* is ubiquitous between 40°N and 40°S and dominates in the oligotrophic (nutrient poor) regions of the oceans. The bacterium accounts for an estimated 20% of the oxygen in the Earth's atmosphere, and forms part of the base of the ocean food chain.

Bacterial Biomass

There are typically 50 million bacterial cells in a gram of soil and a million bacterial cells in a millilitre of fresh water. In a much-cited study from 1998 the world bacterial biomass was calculated to be 350 to 550 billions of tonnes of carbon, equal to between 60% and 100% of the carbon in plants. More recent studies of seafloor microbes have cast considerable doubt on that, one study in 2012 reduced the calculated microbial biomass on the seafloor from the original 303 billions of tonnes of C to just 4.1 billions of tonnes of C, reducing the global biomass of prokaryotes to 50 to 250 billions of tonnes of C. Further, if the average per cell biomass of prokaryotes is reduced from 86 to 14 femtograms C then the global biomass of prokaryotes is reduced to 13 to 44.5 billions of tonnes of C, equal to between 2.4% and 8.1% of the carbon in plants.

Geographic location	Number of cells ($\times 10^{29}$)	Billions of tonnes of carbon
Ocean floor	2.9 to 50	4.1 to 303
Open ocean	1.2	1.7 to 10
Terrestrial soil	2.6	3.7 to 22
Subsurface terrestrial	2.5 to 25	3.5 to 215

Global Biomass

Estimates for the global biomass of species and higher level groups are not always consistent across the literature. Apart from bacteria, the total global biomass has been estimated at about 560 billion tonnes C. Most of this biomass is found on land, with only 5 to 10 billion tonnes C found in the oceans. On land, there is about 1,000 times more plant biomass (*phytomass*) than animal biomass (*zoomass*). About 18% of this plant biomass is eaten by the land animals. However, in the ocean, the animal biomass is nearly 30 times larger than the plant biomass. Most ocean plant biomass is eaten by the ocean animals.

Humans comprise about 100 million tonnes of the Earth's dry biomass, domesticated animals about 700 million tonnes, and crops about 2 billion tonnes. The most successful animal species, in terms of biomass, may well be Antarctic krill, *Euphausia superba*, with a fresh biomass approaching 500 million tonnes, although domestic cattle may also reach these immense figures. However, as a group, the small aquatic crustaceans called copepods may form the largest animal biomass on earth. A 2009 paper in *Science* estimates, for the first time, the total world fish biomass as somewhere between 0.8 and 2.0 billion tonnes. It has been estimated that about 1% of the global biomass is due to phytoplankton, and fully 25% is due to fungi.

Grasses, trees and shrubs have a much higher biomass than the animals that consume them

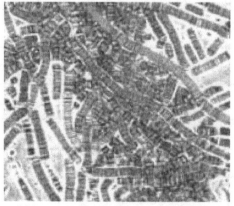

The total biomass of bacteria may equal that of plants.

	name	number of species	date of estimate	individual count	mean living mass of individual	percent biomass (dried)	total number of carbon atoms	global dry biomass in million tonnes	global wet (fresh) biomass in million tonnes
Terrestrial	Humans	1	2012	7.0 billion	50 kg (incl children)	30%	3.5×10^{28}	105	350
			2005	4.63 billion	62 kg (excl children)				287
	Cattle	1		1.3 billion	400 kg	30%		156	520
	Sheep and goats	2	2002	1.75 billion	60 kg	30%		31.5	105
	Chickens	1		24 billion	2 kg	30%		14.4	48
	Ants	12,649		10^7 - 10^8 billion	3×10^{-6} kg (0.003 grams)	30%		10–100	30-300
	Termites	>2,800	1996						445
Marine	Blue whales	1	Pre-whaling	340,000		40%			36
			2001	4700		40%			0.5
	Fish	>10,000	2009					800-2,000	
	Antarctic krill	1	1924–2004	7.8×10^{14}	0.486 g				379
	Copepods (a zooplankton)	13,000			10^{-6} - 10^{-9} kg		1×10^{37}		
	Cyanobacteria (a picoplankton)	?	2003						1,000
Global	Prokaryotes (bacteria)	?	1998	4–6 x 10^{30} cells			1.76-2.76 x 10^{40}	350,000-550,000	

Global Rate of Production

Net primary production is the rate at which new biomass is generated, mainly due to photosynthesis. Global primary production can be estimated from satellite observations. Satellites scan the normalised difference vegetation index (NDVI) over terrestrial habitats, and scan sea-surface chlorophyll levels over oceans. This results in 56.4

billion tonnes C/yr (53.8%), for terrestrial primary production, and 48.5 billion tonnes C/yr for oceanic primary production. Thus, the total photoautotrophic primary production for the Earth is about 104.9 billion tonnes C/yr. This translates to about 426 gC/m²/yr for land production (excluding areas with permanent ice cover), and 140 gC/m²/yr for the oceans.

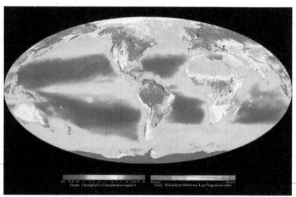

Globally, terrestrial and oceanic habitats produce a similar amount of new biomass each year (56.4 billion tonnes C terrestrial and 48.5 billion tonnes C oceanic).

However, there is a much more significant difference in standing stocks—while accounting for almost half of total annual production, oceanic autotrophs account for only about 0.2% of the total biomass. Autotrophs may have the highest global proportion of biomass, but they are closely rivaled or surpassed by microbes.

Terrestrial freshwater ecosystems generate about 1.5% of the global net primary production.

Some global producers of biomass in order of productivity rates are

Producer	Biomass productivity (gC/m²/yr)	Total area (million km²)	Total production (billion tonnes C/yr)
Swamps and Marshes	2,500		
Tropical rainforests	2,000	8	16
Coral reefs	2,000	0.28	0.56
Algal beds	2,000		
River estuaries	1,800		
Temperate forests	1,250	19	24
Cultivated lands	650	17	11
Tundras	140		
Open ocean	125	311	39
Deserts	3	50	0.15

Organic Matter

Organic matter or organic material, natural organic matter, NOM refers to the large pool of carbon-based compounds found within natural and engineered, terrestrial and aquatic environments. It is matter composed of organic compounds that has come from the remains of organisms such as plants and animals and their waste products in the environment. Organic molecules can also be made by chemical reactions that don't involve life. Basic structures are created from cellulose, tannin, cutin, and lignin, along with other various proteins, lipids, and carbohydrates. Organic matter is very important in the movement of nutrients in the environment and plays a role in water retention on the surface of the planet.

Formation

Living organisms are composed of organic compounds. In life they secrete or excrete organic materials into their environment, shed body parts such as leaves and roots and after the organism dies, its body is broken down by bacterial and fungal action. Larger molecules of organic matter can be formed from the polymerization of different parts of already broken down matter. The composition of natural organic matter depends on its origin, transformation mode, age, and existing environment, thus its bio-physico-chemical functions vary with different environments.

Natural Ecosystem Functions

Organic matter is present throughout the ecosystem. After degrading and reacting, it can move into soil and mainstream water via waterflow. Organic matter provides nutrition to living organisms. Organic matter acts as a buffer in aqueous solution to maintain a neutral pH in the environment. The buffer acting component has been proposed to be relevant for neutralizing acid rain.

Source Cycle

A majority of organic matter not already in the soil comes from groundwater. When the groundwater saturates the soil or sediment around it, organic matter can freely move between the phases. Groundwater has its own sources of natural organic matter also:

- "organic matter deposits, such as kerogen and coal

- soil and sediment organic matter

- organic matter infiltrating into the subsurface from rivers, lakes, and marine systems"

Note that one source of groundwater organic matter is soil organic matter and sedimentary organic matter. The major method of movement into soil is from ground-water, but organic matter from soil moves into groundwater as well. Most of the organic matter in lakes, rivers, and surface water areas comes from deteriorated material in the water and surrounding shores. However, organic matter can pass into or out of water to soil and sediment in the same respect as with the soil.

Importance of the Cycle

Organic matter can migrate through soil, sediment, and water. This movement enables a cycle. Organisms decompose into organic matter, which can then be transported and recycled. Not all biomass migrates, some is rather stationary, turning over only over the course of millions of years.

Soil Organic Matter

The organic matter in soil derives from plants and animals. In a forest, for example, leaf litter and woody material falls to the forest floor. This is sometimes referred to as organic material. When it decays to the point in which it is no longer recognizable it is called soil organic matter. When the organic matter has broken down into a stable substance that resist further decomposition it is called humus. Thus soil organic matter comprises all of the organic matter in the soil exclusive of the material that has not decayed.

One of the advantages of humus is that it is able to withhold water and nutrients, therefore giving plants the capacity for growth. Another advantage of humus is that it helps the soil to stick together which allows nematodes, or microscopic bacteria, to easily decay the nutrients in the soil.

There are several ways to quickly increase the amount of humus. Combining compost, plant or animal materials/waste, or green manure with soil will increase the amount of humus in the soil.

1. Compost: decomposed organic material.

2. Plant and animal material and waste: dead plants or plant waste such as leaves or bush and tree trimmings, or animal manure.

3. Green manure: plants or plant material that is grown for the sole purpose of being incorporated with soil.

These three materials supply nematodes and bacteria with nutrients for them to thrive and produce more humus, which will give plants enough nutrients to survive and grow.

Factors Controlling Rates of Decomposition

o   Environmental factors

- - 1. Aeration

  - 2. Temperature

  - 3. Soil Moisture

  - 4. Soil pH

  o Quality of added residues

  - - 1. Size of organic residues

    - 2. C/N of organic residues

- Rate of decomposition of plant residues, in order from fastest to slowest decomposition rates:

  o 1. Sugars, starches, simple proteins

  o 2. Hemicellulose

  o 3. Cellulose

  o 4. Fats, waxes, oils, resins

  o 5. Lignin, phenolic compounds

Priming Effect

The *priming effect* is characterized by intense changes in the natural process of soil organic matter (SOM) turnover, resulting from relatively moderate intervention with the soil. The phenomenon is generally caused by either pulsed or continuous changes to inputs of fresh organic matter (FOM). Priming effects usually result in an acceleration of mineralization due to a *trigger* such as the FOM inputs. The cause of this increase in decomposition has often been attributed to an increase in microbial activity resulting from higher energy and nutrient availability released from the FOM. After the input of FOM, specialized microorganisms are believed to grow quickly and only decompose this newly added organic matter. The turnover rate of SOM in these areas is at least one order of magnitude higher than the bulk soil.

Other soil treatments, besides organic matter inputs, which lead to this short-term change in turnover rates, include "input of mineral fertilizer, exudation of organic substances by roots, mere mechanical treatment of soil or its drying and rewetting."

Priming effects can be either *positive* or *negative* depending on the reaction of the soil with the added substance. A positive priming effect results in the acceleration of mineralization while a negative priming effect results in immobilization, leading to N unavailability. Although most changes have been documented in C and N pools, the priming effect can also be found in phosphorus and sulfur, as well as other nutrients.

Löhnis was the first to discover the priming effect phenomenon in 1926 through his studies of green manure decomposition and its effects on legume plants in soil. He noticed that when adding fresh organic residues to the soil, it resulted in intensified mineralization by the humus N. It was not until 1953, though, that the term *priming effect* was given by Bingeman in his paper titled, *The effect of the addition of organic materials on the decomposition of an organic soil*. Several other terms had been used before *priming effect* was coined, including priming action, added nitrogen interaction (ANI), extra N and additional N. Despite these early contributions, the concept of the priming effect was widely disregarded until about the 1980s-1990s.

The priming effect has been found in many different studies and is regarded as a common occurrence, appearing in most plant soil systems. However, the mechanisms which lead to the priming effect are more complex then originally thought, and still remain generally misunderstood.

Although there is a lot of uncertainty surrounding the reason for the priming effect, a few *undisputed facts* have emerged from the collection of recent research:

1. The priming effect can arise either instantaneously or very shortly (potentially days or weeks) after the addition of a substance is made to the soil.

2. The priming effect is larger in soils that are rich in C and N as compared to those poor in these nutrients.

3. Real priming effects have not been observed in sterile environments.

4. The size of the priming effect increases as the amount of added treatment to the soil increases.

Recent findings suggest that the same priming effect mechanisms acting in soil systems may also be present in aquatic environments, which suggests a need for broader considerations of this phenomenon in the future.

Decomposition

One suitable definition of organic matter is biological material in the process of decaying or decomposing, such as humus. A closer look at the biological material in the process of decaying reveals so-called organic compounds (biological molecules) in the process of breaking up (disintegrating).

The main processes by which soil molecules disintegrates are by bacterial or fungal enzymatic catalysis. If bacteria or fungi were not present on Earth, the process of decomposition would have proceeded much slower.

Organic Chemistry

Measurements of organic matter generally measure only organic compounds or carbon, and so are only an approximation of the level of once-living or decomposed matter. Some definitions of organic matter likewise only consider "organic matter" to refer to only the carbon content, or organic compounds, and do not consider the origins or decomposition of the matter. In this sense, not all organic compounds are created by living organisms, and living organisms do not only leave behind organic material. A clam's shell, for example, while biotic, does not contain much organic carbon, so may not be considered organic matter in this sense. Conversely, urea is one of many organic compounds that can be synthesized without any biological activity.

Organic matter is heterogeneous and very complex. Generally, organic matter, in terms of weight, is:

- 45-55% carbon
- 35-45% oxygen
- 3-5% hydrogen
- 1-4% nitrogen

The molecular weights of these compounds can vary drastically, depending on if they repolymerize or not, from 200-20,000 amu. Up to one third of the carbon present is in aromatic compounds in which the carbon atoms form usually 6 membered rings. These rings are very stable due to resonance stabilization, so they are difficult to break down. The aromatic rings are also susceptible to electrophilic and nucleophilic attack from other electron-donating or electron-accepting material, which explains the possible polymerization to create larger molecules of organic matter.

There are also reactions that occur with organic matter and other material in the soil to create compounds never seen before. Unfortunately, it is very difficult to characterize these because so little is known about natural organic matter in the first place. Research is currently being done to figure out more about these new compounds and how many of them are being formed.

Organic Matter in Water (Aquatic)

Aquatic organic matter can be further divided into two subsections: dissolved organic matter (DOM) and particulate organic matter (POM). They are typically differentiated by that which can pass through a 0.45 micrometre filter (DOM), and that which cannot (POM).

Detection of Aquatic Organic Matter

Organic matter plays an important role in drinking water and wastewater treatment and recycling, natural aquatic ecosystems, aquaculture, and environmental rehabilitation. It is therefore important to have reliable methods of detection and characterisation, for both short- and long-term monitoring. A variety of analytical detection methods for organic matter have existed for up to decades, to describe and characterise organic matter. These include, but are not limited to: total and dissolved organic carbon, mass spectrometry, nuclear magnetic resonance (NMR) spectroscopy, infrared (IR) spectroscopy, UV-Visible spectroscopy, and fluorescence spectroscopy. Each of these methods has its own advantages and limitations.

Water Purification

The same capability of natural organic matter that helped with water retention in soil creates problems for current water purification methods. In water, organic matter can still bind to metal ions and minerals. These bound molecules are not necessarily stopped by the purification process, but do not cause harm to any humans, animals, or plants. However, because of the high level of reactivity of organic matter, by-products that do not contain nutrients can be made. These by-products can induce biofouling, which essentially clogs water filtration systems in water purification facilities, as the by-products are larger than membrane pore sizes. This clogging problem can be treated by chlorine disinfection (chlorination), which can break down residual material that clogs systems. However, chlorination can form disinfection by-products.

Potential Solutions

Water with organic matter can be disinfected with ozone-initiated radical reactions. The ozone (three oxygens) has very strong oxidation characteristics. It can form hydroxyl radicals (OH) when it decomposes, which will react with the organic matter to shut down the problem of biofouling.

False Positives

Many water quality groups, such as the North Carolina State University Water Quality Group, believe that having too much organic material will cause deoxygenation and essentially remove oxygen from the water. Although organic material, which consists of many hydrocarbon and cyclic carbon chains, is susceptible to attack by oxygen, it would be sterically unfavorable to attach oxygens to every single carbon.

Of course, there are exceptions, such as varying the temperature at which these reactions occur. As the temperature becomes much higher, there is a better chance that an unfavorable reaction will occur because molecules move around faster increasing the randomness of the system (entropy).

Vitalism

The equation of "organic" with living organisms comes from the now-abandoned idea of vitalism that attributed a special force to life that alone could create organic substances. This idea was first questioned after the artificial synthesis of urea by Friedrich Wöhler in 1828.

References

- Groombridge, B.; Jenkins, M. (2002). World Atlas of Biodiversity: Earth's Living Resources in the 21st Century. World Conservation Monitoring Centre, United Nations Environment Programme. ISBN 0-520-23668-8. PMC 3408371.

- Park, Chris C. (2001). The environment: principles and applications (2nd ed.). Routledge. p. 564. ISBN 978-0-415-21770-5.

- Sejian, Veerasamy; Gaughan, John; Baumgard, Lance; Prasad, Cadaba. Climate Change Impact on Livestock: Adaptation and Mitigation. Springer. ISBN 978-81-322-2265-1.

- Nuwer, Rachel (18 July 2015). "Counting All the DNA on Earth". The New York Times. New York: The New York Times Company. ISSN 0362-4331. Retrieved 2015-07-18.

- "Energy crops". crops are grown specifically for use as fuel. BIOMASS Energy Centre. Retrieved 6 April 2013.

- "Biomass for Electricity Generation". capacity of about 6.7 gigawatts in 2000 to about 10.4 gigawatts by 2020. U.S. Energy Information Administration (EIA). Retrieved 6 April 2013.

- "Learning About Renewable Energy". NREL's vision is to develop technology. National Renewable Energy Laboratory. Retrieved 4 April 2013.

- Owning and Operating Costs of Waste and Biomass Power Plants. Claverton-energy.com (2010-09-17). Retrieved on 2012-02-28.

- Biodiesel Will Not Drive Down Global Warming. Energy-daily.com (2007-04-24). Retrieved on 2012-02-28.

- Forest volume-to-biomass models and estimates of mass for live and standing dead trees of U.S. forests. (PDF) . Retrieved on 2012-02-28.

- Scheck, Justin; et al. (July 23, 2012). "Wood-Fired Plants Generate Violations". Wall Street Journal. Retrieved September 27, 2012.

Bioenergy: An Overview

Bioenergy is a renewable source of energy; it stores sunlight in the form of chemical energy. Bioenergy comprises wood, wood waste, manure and other products from agricultural processes. Bioconversion, biomass heating system and biofuels are some aspects of bioenergy that have been discussed in the following section.

Bioenergy

A Stirling engine capable of producing electricity from biomass combustion heat

Bioenergy is renewable energy made available from materials derived from biological sources. Biomass is any organic material which has stored sunlight in the form of chemical energy. As a fuel it may include wood, wood waste, straw, manure, sugarcane, and many other by products from a variety of agricultural processes. By 2010, there was 35 GW (47,000,000 hp) of globally installed bioenergy capacity for electricity generation, of which 7 GW (9,400,000 hp) was in the United States.

In its most narrow sense it is a synonym to biofuel, which is fuel derived from biological sources. In its broader sense it includes biomass, the biological material used as a biofuel, as well as the social, economic, scientific and technical fields associated with using biological sources for energy. This is a common misconception, as bioenergy is the energy extracted from the biomass, as the biomass is the fuel and the bioenergy is the energy contained in the fuel

There is a slight tendency for the word *bioenergy* to be favoured in Europe compared with *biofuel* in America.

Solid Biomass

Simple use of biomass fuel (Combustion of wood for heat).

One of the advantages of biomass fuel is that it is often a by-product, residue or waste-product of other processes, such as farming, animal husbandry and forestry. In theory this means there is no competition between fuel and food production, although this is not always the case. Land use, existing biomass industries and relevant conversion technologies must be considered when evaluating suitability of developing biomass as feedstock for energy.

Biomass is the material derived from recently living organisms, which includes plants, animals and their byproducts. Manure, garden waste and crop residues are all sources of biomass. It is a renewable energy source based on the carbon cycle, unlike other natural resources such as petroleum, coal, and nuclear fuels. Another source includes Animal waste, which is a persistent and unavoidable pollutant produced primarily by the animals housed in industrial-sized farms.

There are also agricultural products specifically being grown for biofuel production. These include corn, and soybeans and to some extent willow and switchgrass on a pre-commercial research level, primarily in the United States; rapeseed, wheat, sugar beet, and willow (15,000 ha or 37,000 acres in Sweden) primarily in Europe; sugarcane in Brazil; palm oil and miscanthus in Southeast Asia; sorghum and cassava in China; and jatropha in India. Hemp has also been proven to work as a biofuel. Biodegradable outputs from industry, agriculture, forestry and households can be used for biofuel production, using e.g. anaerobic digestion to produce biogas, gasification to produce syngas or by direct combustion. Examples of biodegradable wastes include straw, timber, manure, rice husks, sewage, and food waste. The use of biomass fuels can therefore contribute to waste management as well as fuel security and help to prevent or slow down climate change, although alone they are not a comprehensive solution to these problems.

Biomass can be converted to other usable forms of energy like methane gas or transportation fuels like ethanol and biodiesel. Rotting garbage, and agricultural and human waste, all release methane gas—also called "landfill gas" or "biogas." Crops, such as corn and sugar cane, can be fermented to produce the transportation fuel, ethanol. Biodiesel, another transportation fuel, can be produced from left-over food products like vegetable oils and animal fats. Also, Biomass to liquids (BTLs) and cellulosic ethanol are still under research.

Sewage Biomass

A new bioenergy sewage treatment process aimed at developing countries is now on the horizon; the Omni Processor is a self-sustaining process which uses the sewerage solids as fuel to convert sewage waste water into drinking water and electrical energy.

Electricity Generation from Biomass

The biomass used for electricity production ranges by region. Forest byproducts, such as wood residues, are popular in the United States. Agricultural waste is common in Mauritius (sugar cane residue) and Southeast Asia (rice husks). Animal husbandry residues, such as poultry litter, is popular in the UK.

Electricity from Sugarcane Bagasse in Brazil

Sugarcane (Saccharum officinarum) plantation ready for harvest, Ituverava, São Paulo State. Brazil.

Sucrose accounts for little more than 30% of the chemical energy stored in the mature plant; 35% is in the leaves and stem tips, which are left in the fields during harvest, and 35% are in the fibrous material (bagasse) left over from pressing.

The production process of sugar and ethanol in Brazil takes full advantage of the energy stored in sugarcane. Part of the bagasse is currently burned at the mill to provide heat for distillation and electricity to run the machinery. This allows ethanol plants to be energetically self-sufficient and even sell surplus electricity to utilities; current production is 600 MW (800,000 hp) for self-use and 100 MW (130,000 hp) for sale. This secondary activity is expected to boom now that utilities have been induced to pay "fair price "(about US$10/GJ or US$0.036/kWh) for 10 year contracts. This is approximately half of what the World Bank considers the reference price for investing in similar projects. The energy is especially valuable to utilities because it is produced mainly in the dry season when hydroelectric dams are running low. Estimates of potential power generation from bagasse range from 1,000 to 9,000 MW (1,300,000 to 12,100,000 hp), depending on technology. Higher estimates assume gasification of biomass, replacement of current low-pressure steam boilers and turbines by high-pressure ones, and use of harvest trash currently left behind in the fields. For comparison, Brazil's Angra I nuclear plant generates 657 MW (881,000 hp).

A sugar/ethanol plant located in Piracicaba, São Paulo State. This plant produces the electricity it needs from bagasse residuals from sugarcane left over by the milling process, and it sells the surplus electricity to the public grid.

Presently, it is economically viable to extract about 288 MJ of electricity from the residues of one tonne of sugarcane, of which about 180 MJ are used in the plant itself. Thus a medium-size distillery processing 1,000,000 tonnes (980,000 long tons; 1,100,000 short tons) of sugarcane per year could sell about 5 MW (6,700 hp) of surplus electricity. At current prices, it would earn US$18 million from sugar and ethanol sales, and about US$1 million from surplus electricity sales. With advanced boiler and turbine technology, the electricity yield could be increased to 648 MJ per tonne of sugarcane, but current electricity prices do not justify the necessary investment. (According to one report, the World Bank would only finance investments in bagasse power generation if the price were at least US$19/GJ or US$0.068/kWh.)

Bagasse burning is environmentally friendly compared to other fuels like oil and coal. Its ash content is only 2.5% (against 30–50% of coal), and it contains very little sulfur. Since it burns at relatively low temperatures, it produces little nitrous oxides. Moreover, bagasse is being sold for use as a fuel (replacing heavy fuel oil) in various industries, including citrus juice concentrate, vegetable oil, ceramics, and tyre recycling. The state of São Paulo alone used 2,000,000 tonnes (1,970,000 long tons; 2,200,000 short tons), saving about US$35 million in fuel oil imports.

Researchers working with cellulosic ethanol are trying to make the extraction of ethanol from sugarcane bagasse and other plants viable on an industrial scale.

Electricity from Electrogenic Micro-organisms

Another form of bioenergy can be attained from microbial fuel cells, in which chemical energy stored in wastewater or soil is converted directly into electrical energy via the metabolic processes of electrogenic micro-organisms. The power generation capability of this technology has not been economical to date, however, and this technology has found more utility for chemical treatment processes and student education.

Environmental Impact

Some forms of forest bioenergy have recently come under fire from a number of environmental organizations, including Greenpeace and the Natural Resources Defense Council, for the harmful impacts they can have on forests and the climate. Greenpeace recently released a report entitled Fuelling a BioMess which outlines their concerns around forest bioenergy. Because any part of the tree can be burned, the harvesting of trees for energy production encourages Whole-Tree Harvesting, which removes more nutrients and soil cover than regular harvesting, and can be harmful to the long-term health of the forest. In some jurisdictions, forest biomass is increasingly consisting of elements essential to functioning forest ecosystems, including standing trees, naturally disturbed forests and remains of traditional logging operations that were previously left in the forest. Environmental groups also cite recent scientific research which has found that it can take many decades for the carbon released by burning biomass to be recaptured by regrowing trees, and even longer in low productivity areas; further more, logging operations may disturb forest soils and cause them to release stored carbon. In light of the pressing need to reduce greenhouse gas emissions in the short term in order to mitigate the effects of climate change, a number of environmental groups are opposing the large-scale use of forest biomass in energy production.

The New Scientist described a scenario in a September 2016 article which illustrated why the journal believed bioenergy can be bad: Suppose you cut down a 50 year oak tree in your garden and use the logs to heat your house instead of coal. Wood emits more carbon dioxide than coal per unit of heat gained and the roots left in the soil emit more carbon dioxide as they rot. If you plant another tree it will soak up that carbon dioxide in about 50 years. But if you had left the original tree in place it would have soaked up the carbon dioxide from the coal and more. It could take centuries before cutting down the tree would give any benefit. But the world needed to cut carbon dioxide over the next few decades if the global warming was to be kept below 3 degrees C. The journal also concluded that official claimed carbon reductions from renewables had been overstated. The European Union, for example, got more 64% of its renewable energy from biomass (mostly wood) but United Nations and EU rules did not count the carbon emissions from burning biomass.

Bioconversion

Bioconversion, also known as *biotransformation*, is the conversion of organic materials, such as plant or animal waste, into usable products or energy sources by biological processes or agents, such as certain microorganisms. One example is the industrial production of cortisone, which one step is the bioconversion of progesterone to 11-alpha-Hydroxyprogesterone by *Rhizopus nigricans*. Another example is the bioconversion of glycerol to 1,3-propanediol, which is part of scientific research for many decades.

Another example of bioconversion is the conversion of organic materials, such as plant or animal waste, into usable products or energy sources by biological processes or agents, such as certain microorganisms, some detritivores or enzymes.

In the USA, the Bioconversion Science and Technology group performs multidisciplinary R&D for the Department of Energy's (DOE) relevant applications of bioprocessing, especially with biomass. Bioprocessing combines the disciplines of chemical engineering, microbiology and biochemistry. The Group 's primary role is investigation of the use of microorganism, microbial consortia and microbial enzymes in bioenergy research.

New cellulosic ethanol conversion processes have enabled the variety and volume of feedstock that can be bioconverted to expand rapidly. Feedstock now includes materials derived from plant or animal waste such as paper, auto-fluff, tires, fabric, construction materials, municipal solid waste (MSW), sludge, sewage, etc.

Three Different Processes for Bioconversion

1 - Enzymatic hydrolysis - a single source of feedstock, switchgrass for example, is mixed with strong enzymes which convert a portion of cellulosic material into sugars which can then be fermented into ethanol. Genencor and Novozymes are two companies that have received United States government Department of Energy funding for research into reducing the cost of cellulase, a key enzyme in the production cellulosic ethanol by this process.

2 - Synthesis gas fermentation - a blend of feedstock, not exceeding 30% water, is gasified in a closed environment into a syngas containing mostly carbon monoxide and hydrogen. The cooled syngas is then converted into usable products through exposure to bacteria or other catalysts. BRI Energy, LLC is a company whose pilot plant in Fayetteville, Arkansas is currently using synthesis gas fermentation to convert a variety of waste into ethanol. After gasification, anaerobic bacteria (*Clostridium ljungdahlii*) are used to convert the syngas (CO, CO_2, and H_2) into ethanol. The heat generated by gasification is also used to co-generate excess electricity.

3 - C.O.R.S. and Grub Composting are sustainable technologies that employ organisms that feed on organic matter to reduce and convert organic waste in to high quality feedstuff and oil rich material for the biodiesel industry. Organizations pioneering this novel approach to waste management are EAWAG, ESR International, Prota Culture and BIOCONVERSION that created the *e*-CORS® system to meet large scale organic waste management needs and environmental sustainability in both urban and livestock

farming reality. This type of engineered system introduces a substantial innovation represented by the automatic modulation of the treatment, able to adapt conditions of the system to the biology of the scavenger used, improving their performances and the power of this technology.

Biomass Heating System

Wood chips in a storage hopper, in the middle an agitator to transport the material with a screw conveyor to the boiler

Biomass heating systems generate heat from biomass. The systems fall under the categories of:

- direct combustion,

- gasification,

- combined heat and power (CHP),

- anaerobic digestion,

- aerobic digestion.

Benefits of Biomass Heating

The use of biomass in heating systems is beneficial because it uses agricultural, forest, urban and industrial residues and waste to produce heat and electricity with less effect on the environment than fossil fuels. This type of energy production has a limited long-term effect on the environment because the carbon in biomass is part of the natural carbon cycle; while the carbon in fossil fuels is not, and permanently adds carbon to the environment when burned for fuel (carbon footprint). Historically, before the use of fossil fuels in significant quantities, biomass in the form of wood fuel provided most of humanity's heating.

Drawbacks of Biomass Heating

On a large scale, the use of biomass takes agricultural land out of food production, reduces the carbon sequestration capacity of forests, and extracts nutrients from the soil. Combustion of biomass creates air pollutants and adds significant quantities of carbon to the atmosphere that may not be returned to the soil for many decades.

Using biomass as a fuel produces air pollution in the form of carbon monoxide, NOx (nitrogen oxides), VOCs (volatile organic compounds), particulates and other pollutants, in some cases at levels above those from traditional fuel sources such as coal or natural gas. Black carbon – a pollutant created by incomplete combustion of fossil fuels, biofuels, and biomass – is possibly the second largest contributor to global warming. In 2009 a Swedish study of the giant brown haze that periodically covers large areas in South Asia determined that it had been principally produced by biomass burning, and to a lesser extent by fossil-fuel burning. Researchers measured a significant concentration of C, which is associated with recent plant life rather than with fossil fuels.

On combustion, the carbon from biomass is released into the atmosphere as carbon dioxide (CO_2). The amount of carbon stored in dry wood is approximately 50% by weight. When from agricultural sources, plant matter used as a fuel can be replaced by planting for new growth. When the biomass is from forests, the time to recapture the carbon stored is generally longer, and the carbon storage capacity of the forest may be reduced overall if destructive forestry techniques are employed.

The biomass-is-carbon-neutral proposal put forward in the early 1990s has been superseded by more recent science that recognizes that mature, intact forests sequester carbon more effectively than cut-over areas. When a tree's carbon is released into the atmosphere in a single pulse, it contributes to climate change much more than woodland timber rotting slowly over decades. Current studies indicate that "even after 50 years the forest has not recovered to its initial carbon storage" and "the optimal strategy is likely to be protection of the standing forest".

Biomass Heating in Our World

The oil price increases since 2003 and consequent price increases for natural gas and coal have increased the value of biomass for heat generation. Forest renderings, agricultural waste, and crops grown specifically for energy production become competitive as the prices of energy dense fossil fuels rise. Efforts to develop this potential may have the effect of regenerating mismanaged croplands and be a cog in the wheel of a decentralized, multi-dimensional renewable energy industry. Efforts to promote and advance these methods became common throughout the European Union through the 2000s. In other areas of the world, inefficient and polluting means to generate heat from biomass coupled with poor forest practices have significantly added to environmental degradation.

Buffer Tanks

Buffer tanks store the hot water the biomass appliance generates and circulates it around the heating system. Sometimes referred to as 'thermal stores', they are crucial for the efficient operation of all biomass boilers where the system loading fluctuates rapidly, or the volume of water in the complete hydraulic system is relatively small. Using a suitably sized buffer vessel prevents rapid cycling of the boiler when the loading is below the minimum boiler output. Rapid cycling of the boiler causes a large increase in harmful emissions such as Carbon monoxide, dust, and NOx, greatly reduces boiler efficiency and increases electrical consumption of the unit. In addition, service and maintenance requirements will be increased as parts are stressed by rapid heating and cooling cycles. Although most boilers claim to be able to turn down to 30% of nominal output, in the real world this is often not achievable due to differences in the fuel from the 'ideal' or test fuel. A suitably sized buffer tank should therefore be considered where the loading of the boiler drops below 50% of the nominal output – in other words unless the biomass component is purely base load, the system should include a buffer tank. In any case where the secondary system does not contain sufficient water for safe removal of residual heat from the biomass boiler irrespective of the loading conditions, the system must include a suitably sized buffer tank. The residual heat from a biomass unit varies greatly depending on the boiler design and the thermal mass of the combustion chamber. light weight, fast response boilers require only 10ltr/kW, while industrial wet wood units with very high thermal mass require 40ltr/kW.

Types of Biomass Heating Systems

Little biomass heating plant in Austria; the heat power is about 1000 kW

The use of Biomass in heating systems has a use in many different types of buildings, and all have different uses. There are four main types of heating systems that use biomass to heat a boiler. The types are Fully Automated, Semi-Automated, Pellet-Fired, and Combined Heat and Power.

Fully Automated

Fully automated systems operate exactly how they sound. Chipped or ground up waste wood is brought to the site by delivery trucks and dropped into a holding tank. A system of conveyors then transports the wood from the holding tank to the boiler at a certain managed rate. This rate is managed by computer controls and a laser that measures the load of fuel the conveyor is bringing in. The system automatically goes on and off to maintain the pressure and temperature within the boiler. Fully automated systems offer a great deal of ease in their operation because they only require the operator of the system to control the computer, and not the transport of wood while offering comprehensive and cost effective solutions to complex industrial challenges.

Semi-automated or "Surge Bin"

Semi-Automated or "Surge Bin" systems are very similar to fully automated systems except they require more manpower to keep operational. They have smaller holding tanks, and a much simpler conveyor systems which will require personnel to maintain the systems operation. The reasoning for the changes from the fully automated system is the efficiency of the system. The heat created by the combustor can be used to directly heat the air or it can be used to heat water in a boiler system which acts as the medium by which the heat is delivered. Wood fire fueled boilers are most efficient when they are running at their highest capacity, and the heat required most days of the year will not be the peak heat requirement for the year. Considering that the system will only need to run at a high capacity a few days of the year, it is made to meet the requirements for the majority of the year to maintain its high efficiency.

Pellet-fired

The third main type of biomass heating systems are pellet-fired systems. Pellets are a processed form of wood, which make them more expensive. Although they are more expensive, they are much more condensed and uniform, and therefore are more efficient. Further, it is relatively easy to automatically feed pellets to boilers. In these systems, the pellets are stored in a grain-type storage silo, and gravity is used to move them to the boiler. The storage requirements are much smaller for pellet-fired systems because of their condensed nature, which also helps cut down costs. these systems are used for a wide variety of facilities, but they are most efficient and cost effective for places where space for storage and conveyor systems is limited, and where the pellets are made fairly close to the facility.

Agricultural Pellet Systems

Sub category of pellet systems are boilers or burners capable of burning pellet with higher ash rate (paper pellets, hay pellets, straw pellets). One of this kind is PETRO-JET pellet burner with rotating cylindrical burning chamber. In terms of efficiencies advanced pellet boilers can exceed other forms of biomass because of the more stable fuel charataristics. Advanced pellet boilers can even work in condensing mode and cool down combustion gases to 30-40 °C, instead of 120 °C before sent into the flue.

Combined Heat and Power

Combined heat and power systems are very useful systems in which wood waste, such as wood chips, is used to generate power, and heat is created as a byproduct of the power generation system. They have a very high cost because of the high pressure operation. Because of this high pressure operation, the need for a highly trained operator is mandatory, and will raise the cost of operation. Another drawback is that while they produce electricity they will produce heat, and if producing heat is not desirable for certain parts of the year, the addition of a cooling tower is necessary, and will also raise the cost.

There are certain situations where CHP is a good option. Wood product manufacturers would use a combined heat and power system because they have a large supply of waste wood, and a need for both heat and power. Other places where these systems would be optimal are hospitals and prisons, which need energy, and heat for hot water. These systems are sized so that they will produce enough heat to match the average heat load so that no additional heat is needed, and a cooling tower is not needed.

Biofuel

A bus fueled by biodiesel

A biofuel is a fuel that is produced through contemporary biological processes, such as agriculture and anaerobic digestion, rather than a fuel produced by geological processes such as those involved in the formation of fossil fuels, such as coal and petroleum, from prehistoric biological matter. Biofuels can be derived directly from plants, or indirectly from agricultural, commercial, domestic, and/or industrial wastes. Renewable biofuels generally involve contemporary carbon fixation, such as those that occur in plants or microalgae through the process of photosynthesis. Other renewable biofuels are made through the use or conversion of biomass (referring to recently living organisms, most often referring to plants or plant-derived materials). This biomass can be converted to convenient energy-containing substances in three different ways: thermal conversion, chemical conversion, and biochemical conversion. This biomass conversion can result in fuel in solid, liquid, or gas form. This new biomass can also be used directly for biofuels.

Information on pump regarding ethanol fuel blend up to 10%, California

Bioethanol is an alcohol made by fermentation, mostly from carbohydrates produced in sugar or starch crops such as corn, sugarcane, or sweet sorghum. Cellulosic biomass, derived from non-food sources, such as trees and grasses, is also being developed as a feedstock for ethanol production. Ethanol can be used as a fuel for vehicles in its pure form, but it is usually used as a gasoline additive to increase octane and improve vehicle emissions. Bioethanol is widely used in the USA and in Brazil. Current plant design does not provide for converting the lignin portion of plant raw materials to fuel components by fermentation.

Biodiesel can be used as a fuel for vehicles in its pure form, but it is usually used as a diesel additive to reduce levels of particulates, carbon monoxide, and hydrocarbons from diesel-powered vehicles. Biodiesel is produced from oils or fats using transesterification and is the most common biofuel in Europe.

In 2010, worldwide biofuel production reached 105 billion liters (28 billion gallons US), up 17% from 2009, and biofuels provided 2.7% of the world's fuels for road transport. Global ethanol fuel production reached 86 billion liters (23 billion gallons US) in 2010, with the United States and Brazil as the world's top producers, accounting together for 90% of global production. The world's largest biodiesel producer is the European Union, accounting for 53% of all biodiesel production in 2010. As of 2011, mandates for blending biofuels exist in 31 countries at the national level and in 29 states or provinces. The International Energy Agency has a goal for biofuels to meet more than a quarter of world demand for transportation fuels by 2050 to reduce dependence on petroleum and coal. The production of biofuels also led into a flourishing automotive industry, where by 2010, 79% of all cars produced in Brazil were made with a hybrid fuel system of bioethanol and gasoline.

There are various social, economic, environmental and technical issues relating to biofuels production and use, which have been debated in the popular media and scientific journals. These include: the effect of moderating oil prices, the "food vs fuel" debate, poverty reduction potential, carbon emissions levels, sustainable biofuel production, deforestation and soil erosion, loss of biodiversity, impact on water resources, rural social exclusion and injustice, shantytown migration, rural unskilled unemployment, and nitrous oxide (NO2) emissions.

Liquid Fuels for Transportation

Most transportation fuels are liquids, because vehicles usually require high energy density. This occurs naturally in liquids and solids. High energy density can also be provided by an internal combustion engine. These engines require clean-burning fuels. The fuels that are easiest to burn cleanly are typically liquids and gases. Thus, liquids meet the requirements of being both energy-dense and clean-burning. In addition, liquids (and gases) can be pumped, which means handling is easily mechanized, and thus less laborious.

First-generation Biofuels

"First-generation" or conventional biofuels are made from sugar, starch, or vegetable oil.

Ethanol

Neat ethanol on the left (A), gasoline on the right (G) at a filling station in Brazil

Biologically produced alcohols, most commonly ethanol, and less commonly propanol and butanol, are produced by the action of microorganisms and enzymes through the fermentation of sugars or starches (easiest), or cellulose (which is more difficult).

Biobutanol (also called biogasoline) is often claimed to provide a direct replacement for gasoline, because it can be used directly in a gasoline engine.

Ethanol fuel is the most common biofuel worldwide, particularly in Brazil. Alcohol fuels are produced by fermentation of sugars derived from wheat, corn, sugar beets, sugar cane, molasses and any sugar or starch from which alcoholic beverages such as whiskey, can be made (such as potato and fruit waste, etc.). The ethanol production methods used are enzyme digestion (to release sugars from stored starches), fermentation of the sugars, distillation and drying. The distillation process requires significant energy input for heat (sometimes unsustainable natural gas fossil fuel, but cellulosic biomass such as bagasse, the waste left after sugar cane is pressed to extract its juice, is the most common fuel in Brazil, while pellets, wood chips and also waste heat are more common in Europe) Waste steam fuels ethanol factory- where waste heat from the factories also is used in the district heating grid.

U.S. President George W. Bush looks at sugar cane, a source of biofuel, with Brazilian President Luiz Inácio Lula da Silva during a tour on biofuel technology at Petrobras in São Paulo, Brazil, 9 March 2007.

Ethanol can be used in petrol engines as a replacement for gasoline; it can be mixed with gasoline to any percentage. Most existing car petrol engines can run on blends of up to 15% bioethanol with petroleum/gasoline. Ethanol has a smaller energy density than that of gasoline; this means it takes more fuel (volume and mass) to produce the same amount of work. An advantage of ethanol (CH_3CH_2OH) is that it has a higher octane rating than ethanol-free gasoline available at roadside gas stations, which allows an increase of an engine's compression ratio for increased thermal efficiency. In high-altitude (thin air) locations, some states mandate a mix of gasoline and ethanol as a winter oxidizer to reduce atmospheric pollution emissions.

Ethanol is also used to fuel bioethanol fireplaces. As they do not require a chimney and are "flueless", bioethanol fires are extremely useful for newly built homes and apartments without a flue. The downsides to these fireplaces is that their heat output is slightly less than electric heat or gas fires, and precautions must be taken to avoid carbon monoxide poisoning.

Corn-to-ethanol and other food stocks has led to the development of cellulosic ethanol. According to a joint research agenda conducted through the US Department of Energy, the fossil energy ratios (FER) for cellulosic ethanol, corn ethanol, and gasoline are 10.3, 1.36, and 0.81, respectively.

Ethanol has roughly one-third lower energy content per unit of volume compared to gasoline. This is partly counteracted by the better efficiency when using ethanol (in a long-term test of more than 2.1 million km, the BEST project found FFV vehicles to be 1-26 % more energy efficient than petrol cars The BEST project), but the volumetric consumption increases by approximately 30%, so more fuel stops are required.

With current subsidies, ethanol fuel is slightly cheaper per distance traveled in the United States.

Biodiesel

Biodiesel is the most common biofuel in Europe. It is produced from oils or fats using transesterification and is a liquid similar in composition to fossil/mineral diesel. Chemically, it consists mostly of fatty acid methyl (or ethyl) esters (FAMEs). Feedstocks for biodiesel include animal fats, vegetable oils, soy, rapeseed, jatropha, mahua, mustard, flax, sunflower, palm oil, hemp, field pennycress, Pongamia pinnata and algae. Pure biodiesel (B100) currently reduces emissions with up to 60% compared to diesel Second generation B100.

Biodiesel can be used in any diesel engine when mixed with mineral diesel. In some countries, manufacturers cover their diesel engines under warranty for B100 use, although Volkswagen of Germany, for example, asks drivers to check by telephone with the VW environmental services department before switching to B100. B100 may become more viscous at lower temperatures, depending on the feedstock used. In most cases, biodiesel is compatible with diesel engines from 1994 onwards, which use 'Viton' (by DuPont) synthetic rubber in their mechanical fuel injection systems. Note however, that no vehicles are certified for using neat biodiesel before 2014, as there was no emission control protocol available for biodiesel before this date.

Electronically controlled 'common rail' and 'unit injector' type systems from the late 1990s onwards may only use biodiesel blended with conventional diesel fuel. These engines have finely metered and atomized multiple-stage injection systems that are very sensitive to the viscosity of the fuel. Many current-generation diesel engines are made so that they can run on B100 without altering the engine itself, although this depends on the fuel rail design. Since biodiesel is an effective solvent and cleans residues deposited by mineral diesel, engine filters may need to be replaced more often, as the biofuel dissolves old deposits in the fuel tank and pipes. It also effectively cleans the engine combustion chamber of carbon deposits, helping to maintain efficiency. In many European countries, a 5% biodiesel blend is widely used and is available at thousands of gas stations. Biodiesel is also an oxygenated fuel, meaning it contains a reduced amount of carbon and higher hydrogen and oxygen content than fossil diesel. This improves the combustion of biodiesel and reduces the particulate emissions from unburnt carbon. However, using neat biodiesel may increase NOx-emissions Nylund.N-O & Koponen.K. 2013. Fuel and Technology Alternatives for Buses. Overall Energy Efficiency and Emission Performance. IEA Bioenergy Task 46. Possibly the new emission standards Euro VI/EPA 10 will lead to reduced NOx-levels also when using B100.

Biodiesel is also safe to handle and transport because it is non-toxic and biodegradable, and has a high flash point of about 300 °F (148 °C) compared to petroleum diesel fuel, which has a flash point of 125 °F (52 °C).

In the USA, more than 80% of commercial trucks and city buses run on diesel. The emerging US biodiesel market is estimated to have grown 200% from 2004 to 2005. "By the end of 2006 biodiesel production was estimated to increase fourfold [from 2004] to more than" 1 billion US gallons (3,800,000 m³).

In France, biodiesel is incorporated at a rate of 8% in the fuel used by all French diesel vehicles. Avril Group produces under the brand Diester, a fifth of 11 million tons of biodiesel consumed annually by the European Union. It is the leading European producer of biodiesel.

Other Bioalcohols

Methanol is currently produced from natural gas, a non-renewable fossil fuel. In the future it is hoped to be produced from biomass as biomethanol. This is technically feasible, but the production is currently being postponed for concerns of Jacob S. Gibbs and Brinsley Coleberd that the economic viability is still pending. The methanol economy is an alternative to the hydrogen economy, compared to today's hydrogen production from natural gas.

Butanol (C_4H_9OH) is formed by ABE fermentation (acetone, butanol, ethanol) and experimental modifications of the process show potentially high net energy gains with butanol as the only liquid product. Butanol will produce more energy and allegedly can be burned "straight" in existing gasoline engines (without modification to the engine or car), and is less corrosive and less water-soluble than ethanol, and could be distributed via existing infrastructures. DuPont and BP are working together to help develop butanol. E. coli strains have also been successfully engineered to produce butanol by modifying their amino acid metabolism.

Green Diesel

Green diesel is produced through hydrocracking biological oil feedstocks, such as vegetable oils and animal fats. Hydrocracking is a refinery method that uses elevated temperatures and pressure in the presence of a catalyst to break down larger molecules, such as those found in vegetable oils, into shorter hydrocarbon chains used in diesel engines. It may also be called renewable diesel, hydrotreated vegetable oil or hydrogen-derived renewable diesel. Green diesel has the same chemical properties as petroleum-based diesel. It does not require new engines, pipelines or infrastructure to distribute and use, but has not been produced at a cost that is competitive with petroleum. Gasoline versions are also being developed. Green diesel is being developed in Louisiana and Singapore by ConocoPhillips, Neste Oil, Valero, Dynamic Fuels, and Honeywell UOP as well as Preem in Gothenburg, Sweden, creating what is known as Evolution Diesel.

Biofuel Gasoline

In 2013 UK researchers developed a genetically modified strain of Escherichia coli (E.Coli), which could transform glucose into biofuel gasoline that does not need to be blended. Later in 2013 UCLA researchers engineered a new metabolic pathway to bypass glycolysis and increase the rate of conversion of sugars into biofuel, while KAIST researchers developed a strain capable of producing short-chain alkanes, free fatty acids, fatty esters and fatty alcohols through the fatty acyl (acyl carrier protein (ACP)) to fatty acid to fatty acyl-CoA pathway *in vivo*. It is believed that in the future it will be possible to "tweak" the genes to make gasoline from straw or animal manure.

Vegetable Oil

Filtered waste vegetable oil

Walmart's truck fleet logs millions of miles each year, and the company planned to double the fleet's efficiency between 2005 and 2015. This truck is one of 15 based at Walmart's Buckeye, Arizona distribution center that was converted to run on a biofuel made from reclaimed cooking grease produced during food preparation at Walmart stores.

Straight unmodified edible vegetable oil is generally not used as fuel, but lower-quality oil can and has been used for this purpose. Used vegetable oil is increasingly being processed into biodiesel, or (more rarely) cleaned of water and particulates and used as a fuel.

As with 100% biodiesel (B100), to ensure the fuel injectors atomize the vegetable oil in the correct pattern for efficient combustion, vegetable oil fuel must be heated to reduce its viscosity to that of diesel, either by electric coils or heat exchangers. This is easier in warm or temperate climates. MAN B&W Diesel, Wärtsilä, and Deutz AG, as well as a number of smaller companies, such as Elsbett, offer engines that are compatible with straight vegetable oil, without the need for after-market modifications.

Vegetable oil can also be used in many older diesel engines that do not use common rail or unit injection electronic diesel injection systems. Due to the design of the combustion chambers in indirect injection engines, these are the best engines for use with vegetable oil. This system allows the relatively larger oil molecules more time to burn. Some older engines, especially Mercedes, are driven experimentally by enthusiasts without any conversion, a handful of drivers have experienced limited success with earlier pre-"Pumpe Duse" VW TDI engines and other similar engines with direct injection. Several companies, such as Elsbett or Wolf, have developed professional conversion kits and successfully installed hundreds of them over the last decades.

Oils and fats can be hydrogenated to give a diesel substitute. The resulting product is a straight-chain hydrocarbon with a high cetane number, low in aromatics and sulfur and does not contain oxygen. Hydrogenated oils can be blended with diesel in all proportions. They have several advantages over biodiesel, including good performance at low temperatures, no storage stability problems and no susceptibility to microbial attack.

Bioethers

Bioethers (also referred to as fuel ethers or oxygenated fuels) are cost-effective compounds that act as octane rating enhancers."Bioethers are produced by the reaction of reactive iso-olefins, such as iso-butylene, with bioethanol." Bioethers are created by wheat or sugar beet. They also enhance engine performance, whilst significantly reducing engine wear and toxic exhaust emissions. Though bioethers are likely to replace petroethers in the UK, it is highly unlikely they will become a fuel in and of itself due to the low energy density. Greatly reducing the amount of ground-level ozone emissions, they contribute to air quality.

When it comes to transportation fuel there are six ether additives- 1. Dimethyl Ether (DME) 2. Diethyl Ether (DEE) 3. Methyl Teritiary-Butyl Ether (MTBE) 4. Ethyl *ter*-butyl ether (ETBE) 5. *Ter*-amyl methyl ether (TAME) 6. *Ter*-amyl ethyl Ether (TAEE)

The European Fuel Oxygenates Association (aka EFOA) credits Methyl Tertiary-Butyl Ether (MTBE) and Ethyl ter-butyl ether (ETBE) as the most commonly used ethers in fuel to replace lead. Ethers were brought into fuels in Europe in the 1970s to replace the highly toxic compound. Although Europeans still use Bio-ether additives, the US no longer has an oxygenate requirement therefore bio-ethers are no longer used as the main fuel additive.

Biogas

Biogas is methane produced by the process of anaerobic digestion of organic material by anaerobes. It can be produced either from biodegradable waste materials or by the use of energy crops fed into anaerobic digesters to supplement gas yields. The solid byproduct, digestate, can be used as a biofuel or a fertilizer.

Pipes carrying biogas

- Biogas can be recovered from mechanical biological treatment waste processing systems.

Note: Landfill gas, a less clean form of biogas, is produced in landfills through naturally occurring anaerobic digestion. If it escapes into the atmosphere, it is a potential greenhouse gas.

- Farmers can produce biogas from manure from their cattle by using anaerobic digesters.

Syngas

Syngas, a mixture of carbon monoxide, hydrogen and other hydrocarbons, is produced by partial combustion of biomass, that is, combustion with an amount of oxygen that is not sufficient to convert the biomass completely to carbon dioxide and water. Before partial combustion, the biomass is dried, and sometimes pyrolysed. The resulting gas mixture, syngas, is more efficient than direct combustion of the original biofuel; more of the energy contained in the fuel is extracted.

- Syngas may be burned directly in internal combustion engines, turbines or high-temperature fuel cells. The wood gas generator, a wood-fueled gasification reactor, can be connected to an internal combustion engine.

- Syngas can be used to produce methanol, DME and hydrogen, or converted via the Fischer-Tropsch process to produce a diesel substitute, or a mixture of alcohols that can be blended into gasoline. Gasification normally relies on temperatures greater than 700 °C.

- Lower-temperature gasification is desirable when co-producing biochar, but results in syngas polluted with tar.

Solid Biofuels

Examples include wood, sawdust, grass trimmings, domestic refuse, charcoal, agricultural waste, nonfood energy crops, and dried manure.

When raw biomass is already in a suitable form (such as firewood), it can burn directly in a stove or furnace to provide heat or raise steam. When raw biomass is in an inconvenient form (such as sawdust, wood chips, grass, urban waste wood, agricultural residues), the typical process is to densify the biomass. This process includes grinding the raw biomass to an appropriate particulate size (known as hogfuel), which, depending on the densification type, can be from 1 to 3 cm (0.4 to 1.2 in), which is then concentrated into a fuel product. The current processes produce wood pellets, cubes, or pucks. The pellet process is most common in Europe, and is typically a pure wood product. The other types of densification are larger in size compared to a pellet, and are compatible with a broad range of input feedstocks. The resulting densified fuel is easier to transport and feed into thermal generation systems, such as boilers.

Industry has used sawdust, bark and chips for fuel for decades, primary in the pulp and paper industry, and also bagasse (spent sugar cane) fueled boilers in the sugar cane industry. Boilers in the range of 500,000 lb/hr of steam, and larger, are in routine operation, using grate, spreader stoker, suspension burning and fluid bed combustion. Utilities generate power, typically in the range of 5 to 50 MW, using locally available fuel. Other industries have also installed wood waste fueled boilers and dryers in areas with low cost fuel.

One of the advantages of biomass fuel is that it is often a byproduct, residue or waste-product of other processes, such as farming, animal husbandry and forestry. In theory, this means fuel and food production do not compete for resources, although this is not always the case.

A problem with the combustion of raw biomass is that it emits considerable amounts of pollutants, such as particulates and polycyclic aromatic hydrocarbons. Even modern pellet boilers generate much more pollutants than oil or natural gas boilers. Pellets made from agricultural residues are usually worse than wood pellets, producing much larger emissions of dioxins and chlorophenols.

In spite of the above noted study, numerous studies have shown biomass fuels have significantly less impact on the environment than fossil based fuels. Of note is the US Department of Energy Laboratory, operated by Midwest Research Institute Biomass Power and Conventional Fossil Systems with and without CO2 Sequestration – Comparing the Energy Balance, Greenhouse Gas Emissions and Economics Study. Power generation emits significant amounts of greenhouse gases (GHGs), mainly carbon dioxide (CO_2). Sequestering CO_2 from the power plant flue gas can significantly reduce the GHGs from the power plant itself, but this is not the total picture. CO_2 capture and sequestration consumes additional energy, thus lowering the plant's fuel-to-electricity efficiency. To compensate for this, more fossil fuel must be procured and consumed to make up for lost capacity.

Taking this into consideration, the global warming potential (GWP), which is a combination of CO_2, methane (CH_4), and nitrous oxide (N_2O) emissions, and energy balance of the system need to be examined using a life cycle assessment. This takes into account the upstream processes which remain constant after CO_2 sequestration, as well as the steps required for additional power generation. Firing biomass instead of coal led to a 148% reduction in GWP.

A derivative of solid biofuel is biochar, which is produced by biomass pyrolysis. Biochar made from agricultural waste can substitute for wood charcoal. As wood stock becomes scarce, this alternative is gaining ground. In eastern Democratic Republic of Congo, for example, biomass briquettes are being marketed as an alternative to charcoal to protect Virunga National Park from deforestation associated with charcoal production.

Second-generation (Advanced) Biofuels

Second generation biofuels, also known as advanced biofuels, are fuels that can be manufactured from various types of biomass. Biomass is a wide-ranging term meaning any source of organic carbon that is renewed rapidly as part of the carbon cycle. Biomass is derived from plant materials but can also include animal materials.

First generation biofuels are made from the sugars and vegetable oils found in arable crops, which can be easily extracted using conventional technology. In comparison, second generation biofuels are made from lignocellulosic biomass or woody crops, agricultural residues or waste, which makes it harder to extract the required fuel. A series of physical and chemical treatments might be required to convert lignocellulosic biomass to liquid fuels suitable for transportation.

Sustainable Biofuels

Biofuels in the form of liquid fuels derived from plant materials, are entering the market, driven mainly by the perception that they reduce climate gas emissions, and also by factors such as oil price spikes and the need for increased energy security. However, many of the biofuels that are currently being supplied have been criticised for their adverse impacts on the natural environment, food security, and land use. In 2008, the Nobel-prize winning chemist Paul J. Crutzen published findings that the release of nitrous oxide (N_2O) emissions in the production of biofuels means that overall they contribute more to global warming than the fossil fuels they replace.

The challenge is to support biofuel development, including the development of new cellulosic technologies, with responsible policies and economic instruments to help ensure that biofuel commercialization is sustainable. Responsible commercialization of biofuels represents an opportunity to enhance sustainable economic prospects in Africa, Latin America and Asia.

According to the Rocky Mountain Institute, sound biofuel production practices would not hamper food and fibre production, nor cause water or environmental problems, and would enhance soil fertility. The selection of land on which to grow the feedstocks is a critical component of the ability of biofuels to deliver sustainable solutions. A key consideration is the minimisation of biofuel competition for prime cropland.

Biofuels by Region

There are international organizations such as IEA Bioenergy, established in 1978 by the OECD International Energy Agency (IEA), with the aim of improving cooperation and information exchange between countries that have national programs in bioenergy research, development and deployment. The UN International Biofuels Forum is formed by Brazil, China, India, Pakistan, South Africa, the United States and the European Commission. The world leaders in biofuel development and use are Brazil, the United States, France, Sweden and Germany. Russia also has 22% of world's forest, and is a big biomass (solid biofuels) supplier. In 2010, Russian pulp and paper maker, Vyborgskaya

Cellulose, said they would be producing pellets that can be used in heat and electricity generation from its plant in Vyborg by the end of the year. The plant will eventually produce about 900,000 tons of pellets per year, making it the largest in the world once operational.

Bio Diesel Powered Fast Attack Craft Of Indian Navy patrolling during IFR 2016. The green bands on the vessels are indicative of the fact that the vessels are powered by bio-diesel

Biofuels currently make up 3.1% of the total road transport fuel in the UK or 1,440 million litres. By 2020, 10% of the energy used in UK road and rail transport must come from renewable sources – this is the equivalent of replacing 4.3 million tonnes of fossil oil each year. Conventional biofuels are likely to produce between 3.7 and 6.6% of the energy needed in road and rail transport, while advanced biofuels could meet up to 4.3% of the UK's renewable transport fuel target by 2020.

Air Pollution

Biofuels are different from fossil fuels in regard to greenhouse gases but are similar to fossil fuels in that biofuels contribute to air pollution. Burning produces airborne carbon particulates, carbon monoxide and nitrous oxides. The WHO estimates 3.7 million premature deaths worldwide in 2012 due to air pollution. Brazil burns significant amounts of ethanol biofuel. Gas chromatograph studies were performed of ambient air in São Paulo, Brazil, and compared to Osaka, Japan, which does not burn ethanol fuel. Atmospheric Formaldehyde was 160% higher in Brazil, and Acetaldehyde was 260% higher.

Debates Regarding the Production and Use of Biofuel

There are various social, economic, environmental and technical issues with biofuel production and use, which have been discussed in the popular media and scientific journals. These include: the effect of moderating oil prices, the "food vs fuel" debate, food prices, poverty reduction potential, energy ratio, energy requirements, carbon emissions levels, sustainable biofuel production, deforestation and soil erosion, loss of biodiversity, impact on water resources, the possible modifications necessary to run the engine on biofuel, as well as energy balance and efficiency. The International Resource Panel, which provides independent scientific assessments and expert advice on a variety of resource-related themes, assessed the issues relating to biofuel use in its first report *Towards sustainable production and use of resources: Assessing Biofuels*. "Assessing Biofuels" outlined the wider and interrelated factors that need to be considered when deciding on the relative merits of pursuing one biofuel over another. It concluded that not all biofuels perform equally in terms of their impact on climate, energy security and ecosystems, and suggested that environmental and social impacts need to be assessed throughout the entire life-cycle.

Another issue with biofuel use and production is the US has changed mandates many times because the production has been taking longer than expected. The Renewable Fuel Standard (RFS) set by congress for 2010 was pushed back to at best 2012 to produce 100 million gallons of pure ethanol (not blended with a fossil fuel).

Current Research

Research is ongoing into finding more suitable biofuel crops and improving the oil yields of these crops. Using the current yields, vast amounts of land and fresh water would be needed to produce enough oil to completely replace fossil fuel usage. It would require twice the land area of the US to be devoted to soybean production, or two-thirds to be devoted to rapeseed production, to meet current US heating and transportation needs.

Specially bred mustard varieties can produce reasonably high oil yields and are very useful in crop rotation with cereals, and have the added benefit that the meal left over after the oil has been pressed out can act as an effective and biodegradable pesticide.

The NFESC, with Santa Barbara-based Biodiesel Industries, is working to develop biofuels technologies for the US navy and military, one of the largest diesel fuel users in the world. A group of Spanish developers working for a company called Ecofasa announced a new biofuel made from trash. The fuel is created from general urban waste which is treated by bacteria to produce fatty acids, which can be used to make biofuels.

Ethanol Biofuels

As the primary source of biofuels in North America, many organizations are conducting research in the area of ethanol production. The National Corn-to-Ethanol Research Center (NCERC) is a research division of Southern Illinois University Edwardsville dedicated solely to ethanol-based biofuel research projects. On the federal level, the USDA conducts a large amount of research regarding ethanol production in the United States. Much of this research is targeted toward the effect of ethanol production on domestic food markets. A division of the U.S. Department of Energy, the National Renewable Energy Laboratory (NREL), has also conducted various ethanol research projects, mainly in the area of cellulosic ethanol.

Cellulosic ethanol commercialization is the process of building an industry out of methods of turning cellulose-containing organic matter into fuel. Companies, such as Iogen, POET, and Abengoa, are building refineries that can process biomass and turn it into bioethanol. Companies, such as Diversa, Novozymes, and Dyadic, are producing enzymes that could enable a cellulosic ethanol future. The shift from food crop feedstocks to waste residues and native grasses offers significant opportunities for a range of players, from farmers to biotechnology firms, and from project developers to investors.

As of 2013, the first commercial-scale plants to produce cellulosic biofuels have begun operating. Multiple pathways for the conversion of different biofuel feedstocks are being used. In the next few years, the cost data of these technologies operating at commercial scale, and their relative performance, will become available. Lessons learnt will lower the costs of the industrial processes involved.

In parts of Asia and Africa where drylands prevail, sweet sorghum is being investigated as a potential source of food, feed and fuel combined. The crop is particularly suitable for growing in arid conditions, as it only extracts one seventh of the water used by sugarcane. In India, and other places, sweet sorghum stalks are used to produce biofuel by squeezing the juice and then fermenting into ethanol.

A study by researchers at the International Crops Research Institute for the Semi-Arid Tropics (ICRISAT) found that growing sweet sorghum instead of grain sorghum could increase farmers incomes by US$40 per hectare per crop because it can provide fuel in addition to food and animal feed. With grain sorghum currently grown on over 11 million hectares (ha) in Asia and on 23.4 million ha in Africa, a switch to sweet sorghum could have a considerable economic impact.

Algae Biofuels

From 1978 to 1996, the US NREL experimented with using algae as a biofuels source in the "Aquatic Species Program". A self-published article by Michael Briggs, at the UNH Biofuels Group, offers estimates for the realistic replacement of all vehicular fuel with biofuels by using algae that have a natural oil content greater than 50%, which Briggs suggests can be grown on algae ponds at wastewater treatment plants. This oil-rich algae can then be extracted from the system and processed into biofuels, with the dried remainder further reprocessed to create ethanol. The production of algae to harvest oil for biofuels has not yet been undertaken on a commercial scale, but feasibility studies have been conducted to arrive at the above yield estimate. In addition to its projected high yield, algaculture — unlike crop-based biofuels — does not entail a decrease in food production, since it requires neither farmland nor fresh water. Many companies are pursuing algae bioreactors for various purposes, including scaling up biofuels production to commercial levels. Prof. Rodrigo E. Teixeira from the University of Alabama in Huntsville demonstrated the extraction of biofuels lipids from wet algae using a simple and economical reaction in ionic liquids.

Jatropha

Several groups in various sectors are conducting research on Jatropha curcas, a poisonous shrub-like tree that produces seeds considered by many to be a viable source of biofuels feedstock oil. Much of this research focuses on improving the overall per acre oil yield of Jatropha through advancements in genetics, soil science, and horticultural practices.

SG Biofuels, a San Diego-based jatropha developer, has used molecular breeding and biotechnology to produce elite hybrid seeds that show significant yield improvements over first-generation varieties. SG Biofuels also claims additional benefits have arisen from such strains, including improved flowering synchronicity, higher resistance to pests and diseases, and increased cold-weather tolerance.

Plant Research International, a department of the Wageningen University and Research Centre in the Netherlands, maintains an ongoing Jatropha Evaluation Project that examines the feasibility of large-scale jatropha cultivation through field and laboratory experiments. The Center for Sustainable Energy Farming (CfSEF) is a Los Angeles-based nonprofit research organization dedicated to jatropha research in the areas of plant science, agronomy, and horticulture. Successful exploration of these disciplines is projected to increase jatropha farm production yields by 200-300% in the next 10 years.

Fungi

A group at the Russian Academy of Sciences in Moscow, in a 2008 paper, stated they had isolated large amounts of lipids from single-celled fungi and turned it into biofuels in an economically efficient manner. More research on this fungal species, Cunninghamella japonica, and others, is likely to appear in the near future. The recent discovery of a variant of the fungus Gliocladium roseum (later renamed Ascocoryne sarcoides) points toward the production of so-called myco-diesel from cellulose. This organism was recently discovered in the rainforests of northern Patagonia, and has the unique capability of converting cellulose into medium-length hydrocarbons typically found in diesel fuel. Many other fungi that can degrade cellulose and other polymers have been observed to produce molecules that are currently being engineered using organisms from other kingdoms, suggesting that fungi may play a large role in the bio-production of fuels in the future (reviewed in).

Animal Gut Bacteria

Microbial gastrointestinal flora in a variety of animals have shown potential for the production of biofuels. Recent research has shown that TU-103, a strain of Clostridium bacteria found in Zebra feces, can convert nearly any form of cellulose into butanol fuel. Microbes in panda waste are being investigated for their use in creating biofuels from bamboo and other plant materials. There has also been substantial research into the technology of using the gut microbiomes of wood-feeding insects for the conversion of lignocellulotic material into biofuel.

Greenhouse Gas Emissions

Some scientists have expressed concerns about land-use change in response to greater demand for crops to use for biofuel and the subsequent carbon emissions. The payback period, that is, the time it will take biofuels to pay back the carbon debt they acquire due to land-use change, has been estimated to be between 100 and 1000 years, depending on the specific instance and location of land-use change. However, no-till practices combined with cover-crop practices can reduce the payback period to three years for grassland conversion and 14 years for forest conversion.

A study conducted in the Tocantis State, in northern Brazil, found that many families were cutting down forests in order to produce two conglomerates of oilseed plants, the J. curcas (JC group) and the R. communis (RC group). This region is composed of 15% Amazonian rainforest with high biodiversity, and 80% Cerrado forest with lower biodiversity. During the study, the farmers that planted the JC group released over 2193 Mg CO_2, while losing 53-105 Mg CO_2 sequestration from deforestation; and the RC group farmers released 562 Mg CO_2, while losing 48-90 Mg CO_2 to be sequestered from forest depletion. The production of these types of biofuels not only led into an increased emission of carbon dioxide, but also to lower efficiency of forests to absorb the gases that these farms were emitting. This has to do with the amount of fossil fuel the production of fuel crops involves. In addition, the intensive use of monocropping agriculture requires large amounts of water irrigation, as well as of fertilizers, herbicides and pesticides. This does not only lead to poisonous chemicals to disperse on water runoff, but also to the emission of nitrous oxide (NO_2) as a fertilizer byproduct, which is three hundred times more efficient in producing a greenhouse effect than carbon dioxide (CO_2).

Converting rainforests, peatlands, savannas, or grasslands to produce food crop–based biofuels in Brazil, Southeast Asia, and the United States creates a "biofuel carbon debt" by releasing 17 to 420 times more CO_2 than the annual greenhouse gas (GHG) reductions that these biofuels would provide by displacing fossil fuels. Biofuels made from waste biomass or from biomass grown on abandoned agricultural lands incur little to no carbon debt.

Water Use

In addition to water required to grow crops, biofuel facilities require significant process water.

Various Forms of Biofuel

Grassoline

Grassoline is a biofuel derived from plant material that is used as an alternative to ethanol. The term was coined by Matthew Scoggins, a graduate students of Bruce Dale, in 1991, to capture the idea of taking plant material and converting it into oil.

Grassoline can be made from switchgrass, which is more efficient than other energy crops such as corn, since it does not require as much water and can store sufficient amount of energy to be efficient. Unlike ethanol made from corn, grassoline can be made from the waste product of most plants. Grassoline generally receives criticism due to its efficiency and price of the fuel. Instead of producing cellulosic ethanol via fermentation from edible cereals(a grass from the monocot family) like corn, new techniques could produce grassoline from otherwise unused biomass. Cellulosic biomass(biofuel derived from living or recently living organisms) from wood residues, such as sawdust and construction debris and agricultural wastes, such as corn stalks and all the inedible stalks of other plants can be converted to any type of ethanol fuel, gasoline, diesel, and all the materials that derive from crude oil and its refinery. Energy crops such as switch grass and miscanthus provide high energy content and does not replace food like corn for energy. Miscanthus, however, does not self-fertilize and is permeable to frost, thus it may not serve as appropriate depending on the location.

Charris Ford is the founder of Grassolean Solutions and owns the web domains Grassoline.com and Grassolean.com, since 2001.

Production

According to a study done by the U.S. Department of Agriculture and the Department of Energy, the United States can produce at least 1.3 billion tons of cellulosic biomass each year without decreasing the amount of biomass needed for our food, animal feed, or exports.

Catalytic Fast Pyrolysis

Catalytic Fast Pyrolysis is a fast process in which the Cellulose is broken down to grassoline. In the Catalynic approach the cellulose is heated to 500 degrees celsius in less than one second in a chamber to break apart the oxygen molecules. The catalyst forms chemical reactions that remove oxygen bonds and form carbon rings. After the reaction takes place gasoline is formed along with water, carbon dioxide, and carbon monoxide.

AFEX Treatment

The Ammonia Fiber Expansion(AFEX) pre-treat process, hot concentrated 15 M ammonia is used to breaks down sugar molecules, cellulose and hemicellulose significantly more efficiently than enzymes. After, the rapid pressure release cools and ends the treatment. The result is minor biomass degradation with high yields The process was patented by Bruce Dale, Michigan State University professor. Unlike other process, AFEX is generally done in one step making it more efficient than other processes.

AFEX Process Conditions

- Pressure: 20-30 atm

- Temperature: 70-140 C

- Residence time: 5–10 minutes

- Ammonia: biomass loading: 0.3 - 2.0 to 1 w/w

- Water: biomass content: 0.2 – 2.5 to 1 w/w

Regional Biomass Processing Center

Regional Biomass Processing Center is a conceptual place where the AFEX treated biomass can go to biorefineries, farms and forests, and animal feeders. This will improve the value of cellulosic biomass for animals and biofuel production. This will reduce the density of the biomass for easier transport, simplify contract issues, and increase the land use for biofuels

Potential Energy Grasses

Plant material is cheaper than oil on both energy and mass basis and certain plant material have potential to be energy grasses.

Switchgrass

Switchgrass is a bunch grass native to North America that grow naturally under warm weather with wide adaptation capability and easy germination, allowing the switchgrass to grow quicker; however, it has a low relative yield compared to other energy crops

Sorghum

Sorghum are cultivated in warmer climates, mostly in the tropical regions. Sorghum has the pottential to be an energy grass because it requires low water usage and can make a large yield. Sorghum, however, has an annual cultivation and is difficult to establish to an area and requires a lot of input from fertilizers and pesticides.

Miscanthus

Miscanthus are native to the tropical regions of Africa and Southern Asia. Miscanthus can grow up to 3.5 meters and has been trialed as a biofuel since the 1980s. The benefits of using miscanthus is it can live more than two years and requires low input eliminating the need for extra irrigation, fertilizer and pesticides. The problems with miscanthus arise from the time it takes to establish to an area.

Cost of Change

The cost for petroleum to change to grassoline, would depend on how fast the use of grassoline grows. Change will also be needed in automobiles to be compatible with grassoline. UC Berkeley's Somerville(professor of alternative energy) estimates that a large investment over $325 billion will be needed to build biofactories that can produce the 65 billion gallons of biofuel needed to meet 2030 national goals.

Criticism

Grassoline is highly debated between the public. Some speculate that plant material is needed for fertilization for next harvest. Common misconceptions concerning grassoline is the fuel could have been used as food for livestock, transportation of the material may be counter productive, and the price of biofuels cost more than gas. Farmers also would not get as much money if they supply biomass rather than food since the cultivation of biomass depends on the location and climate and a large investment over $250 million is needed for a cellulosic ethanol biorefinery. Supporters of grassoline believe it could suppress air pollution, eliminate the need to import energy, eliminating the need of high compression engines used for ethanol fuel and more jobs since the work would be obsolete

Biomass Briquettes

Briquette made by a Ruf briquetter out of hay

Biomass briquettes are a biofuel substitute to coal and charcoal. Briquettes are mostly used in the developing world, where cooking fuels are not as easily available. There has been a move to the use of briquettes in the developed world, where they are used to heat industrial boilers in order to produce electricity from steam. The briquettes are cofired with coal in order to create the heat supplied to the boiler.

Ogatan, Japanese charcoal briquettes made from sawdust briquettes *(Ogalite)*.

Quick Grill Briquette made from coconut shell

Composition and Production

Biomass briquettes, mostly made of green waste and other organic materials, are commonly used for electricity generation, heat, and cooking fuel. These compressed compounds contain various organic materials, including rice husk, bagasse, ground nut shells, municipal solid waste, agricultural waste. The composition of the briquettes varies by area due to the availability of raw materials. The raw materials are gathered and compressed into briquette in order to burn longer and make transportation of the goods easier. These briquettes are very different from charcoal because they do not have large concentrations of carbonaceous substances and added materials. Compared to fossil fuels, the briquettes produce low net total greenhouse gas emissions because the materials used are already a part of the carbon cycle.

One of the most common variables of the biomass briquette production process is the way the biomass is dried out. Manufacturers can use torrefaction, carbonization, or varying degrees of pyrolysis. Researchers concluded that torrefaction and carbonization are the most efficient forms of drying out biomass, but the use of the briquette determines which method should be used.

Compaction is another factor affecting production. Some materials burn more efficiently if compacted at low pressures, such as corn stover grind. Other materials such as wheat and barley-straw require high amounts of pressure to produce heat. There are also different press technologies that can be used. A piston press is used to create solid briquettes for a wide array of purposes. Screw extrusion is used to compact biomass into loose, homogeneous briquettes that are substituted for coal in cofiring. This technology creates a toroidal, or doughnut-like, briquette. The hole in the center of the briquette allows for a larger surface area, creating a higher combustion rate.

History

People have been using biomass briquettes in Nepal since before recorded history. Though inefficient, the burning of loose biomass created enough heat for cooking purposes and keeping warm. The first commercial production plant was created in 1982 and produced almost 900 metric tons of biomass. In 1984, factories were constructed that incorporated vast improvements on efficiency and the quality of briquettes. They used a combination of rice husks and molasses. The King Mahendra Trust for Nature Conservation (KMTNC) along with the Institute for Himalayan Conservation (IHC) created a mixture of coal and biomass in 2000 using a unique rolling machine.

Japanese Ogalite

In 1925, Japan independently started developing technology to harness the energy from sawdust briquettes, known as *"Ogalite"*. Between 1964 and 1969, Japan increased production fourfold by incorporating screw press and piston press technology. The member enterprise of 830 or more existed in the 1960s. The new compaction techniques incorporated in these machines made briquettes of higher quality than those in Europe. As a result, European countries bought the licensing agreements and now manufacture Japanese designed machines.

Cofiring

Cofiring relates to the combustion of two different types of materials. The process is primarily used to decrease CO_2 emissions despite the resulting lower energy efficiency and higher variable cost. The combination of materials usually contains a high carbon emitting substance such as coal and a lesser CO_2 emitting material such as biomass. Even though CO_2 will still be emitted through the combustion of biomass, the net carbon emitted is nearly negligible. This is due to the fact that the material gathered for the composition of the briquettes are still contained in the carbon cycle whereas fossil fuel combustion releases CO_2 that has been sequestered for millennia. Boilers in power plants are traditionally heated by the combustion of coal, but if cofiring were to be implemented, then the CO_2 emissions would decrease while still maintaining the heat inputted to the boiler. Implementing cofiring would require few modifications to the current characteristics to power plants, as only the fuel for the boiler would be altered. A moderate investment would be required for implementing biomass briquettes into the combustion process.

Cofiring is considered the most cost-efficient means of biomass. A higher combustion rate will occur when cofiring is implemented in a boiler when compared to burning only biomass. The compressed biomass is also much easier to transport since it is more dense, therefore allowing more biomass to be transported per shipment when compared to loose biomass. Some sources agree that a near-term solution for the greenhouse gas emission problem may lie in cofiring.

Compared to Coal

The use of biomass briquettes has been steadily increasing as industries realize the benefits of decreasing pollution through the use of biomass briquettes. Briquettes provide higher calorific value per dollar than coal when used for firing industrial boilers. Along with higher calorific value, biomass briquettes on average saved 30–40% of boiler fuel cost. But other sources suggest that cofiring is more expensive due to the widespread availability of coal and its low cost. However, in the long run, briquettes can only limit the use of coal to a small extent, but it is increasingly being pursued by industries and factories all over the world. Both raw materials can be produced or mined domestically in the United States, creating a fuel source that is free from foreign dependence and less polluting than raw fossil fuel incineration.

Environmentally, the use of biomass briquettes produces much fewer greenhouse gases, specifically, 13.8% to 41.7% CO_2 and NO_x. There was also a reduction from 11.1% to 38.5% in SO 2 emissions when compared to coal from three different leading producers, EKCC Coal, Decanter Coal, and Alden Coal. Biomass briquettes are also fairly resistant to water degradation, an improvement over the difficulties encountered with the burning of wet coal. However, the briquettes are best used only as a supplement to coal. The use of cofiring creates an energy that is not as high as pure coal, but emits fewer pollutants and cuts down on the release of previously sequestered carbon. The continuous release of carbon and other greenhouse gasses into the atmosphere leads to an increase in global temperatures. The use of cofiring does not stop this process but decreases the relative emissions of coal power plants.

Use in Developing World

The Legacy Foundation has developed a set of techniques to produce biomass briquettes through artisanal production in rural villages that can be used for heating and cooking. These techniques were recently pioneered by Virunga National Park in eastern Democratic Republic of Congo, following the massive destruction of the mountain gorilla habitat for charcoal.

Pangani, Tanzania, is an area covered in coconut groves. After harvesting the meat of the coconut, the indigenous people would litter the ground with the husks, believing them to be useless. The husks later became a profit center after it was discovered that coconut husks are well suited to be the main ingredient in bio briquettes. This alternative fuel mixture burns incredibly efficiently and leaves little residue, making it a reliable source for cooking in the undeveloped country. The developing world has always relied on the burning biomass due it its low cost and availability anywhere there is organic material. The briquette production only improves upon the ancient practice by increasing the efficiency of pyrolysis.

Two major components of the developing world are China and India. The economies are rapidly increasing due to cheap ways of harnessing electricity and emitting large amounts of carbon dioxide. The Kyoto Protocol attempted to regulate the emissions of the three different worlds, but there were disagreements as to which country should be penalized for emissions based on its previous and future emissions. The United States has been the largest emitter but China has recently become the largest per capita. The

United States had emitted a rigorous amount of carbon dioxide during its development and the developing nations argue that they should not be forced to meet the requirements. At the lower end, the undeveloped nations believe that they have little responsibility for what has been done to the carbon dioxide levels. The major use of biomass briquettes in India, is in industrial applications usually to produce steam. A lot of conversions of boilers from FO to biomass briquettes have happened over the past decade. A vast majority of those projects are registered under CDM (Kyoto Protocol), which allows for users to get carbon credits.

The use of biomass briquettes is strongly encouraged by issuing carbon credits. One carbon credit is equal to one free ton of carbon dioxide to be emitted into the atmosphere. India has started to replace charcoal with biomass briquettes in regards to boiler fuel, especially in the southern parts of the country because the biomass briquettes can be created domestically, depending on the availability of land. Therefore, constantly rising fuel prices will be less influential in an economy if sources of fuel can be easily produced domestically. Lehra Fuel Tech Pvt Ltd is approved by Indian Renewable Energy Development Agency (IREDA), is one of the largest briquetting machine manufacturers from Ludhiana, India.

In the African Great Lakes region, work on biomass briquette production has been spearheaded by a number of NGOs with GVEP (Global Village Energy Partnership) taking a lead in promoting briquette products and briquette entrepreneurs in the three Great Lakes countries; namely, Kenya, Uganda and Tanzania. This has been achieved by a five-year EU and Dutch government sponsored project called DEEP EA (Developing Energy Enterprises Project East Africa) . The main feed stock for briquettes in the East African region has mainly been charcoal dust although alternative like sawdust, bagasse, coffee husks and rice husks have also been used.

Use in Developed World

Coal is the largest carbon dioxide emitter per unit area when it comes to electricity generation. It is also the most common ingredient in charcoal. There has been a recent push to replace the burning of fossil fuels with biomass. The replacement of this nonrenewable resource with biological waste would lower the carbon footprint of grill owners and lower the overall pollution of the world. Citizens are also starting to manufacture briquettes at home. The first machines would create briquettes for homeowners out of compressed sawdust, however, current machines allow for briquette production out of any sort of dried biomass.

Arizona has also taken initiative to turn waste biomass into a source of energy. Waste cotton and pecan material used to provide a nesting ground for bugs that would destroy the new crops in the spring. To stop this problem farmers buried the biomass, which quickly led to soil degradation. These materials were discovered to be a very efficient source of energy and took care of issues that had plagued farms.

The United States Department of Energy has financed several projects to test the viability of biomass briquettes on a national scale. The scope of the projects is to increase the efficiency of gasifiers as well as produce plans for production facilities.

Criticism

Biomass is composed of organic materials, therefore, large amounts of land are required to produce the fuel. Critics argue that the use of this land should be utilized for food distribution rather than crop degradation. Also, climate changes may cause a harsh season, where the material extracted will need to be swapped for food rather than energy. The assumption is that the production of biomass decreases the food supply, causing an increase in world hunger by extracting the organic materials such as corn and soybeans for fuel rather than food.

The cost of implementing a new technology such as biomass into the current infrastructure is also high. The fixed costs with the production of biomass briquettes are high due to the new undeveloped technologies that revolve around the extraction, production and storage of the biomass. Technologies regarding extraction of oil and coal have been developing for decades, becoming more efficient with each year. A new undeveloped technology regarding fuel utilization that has no infrastructure built around makes it nearly impossible to compete in the current market.

Biodiesel

Space-filling model of methyl linoleate, or linoleic acid methyl ester, a common methyl ester produced from soybean or canola oil and methanol

Space-filling model of ethyl stearate, or stearic acid ethyl ester, an ethyl ester produced from soybean or canola oil and ethanol

Biodiesel refers to a vegetable oil - or animal fat-based diesel fuel consisting of long-chain alkyl (methyl, ethyl, or propyl) esters. Biodiesel is typically made by chemically reacting lipids (e.g., vegetable oil, soybean oil, animal fat (tallow)) with an alcohol producing fatty acid esters.

Biodiesel is meant to be used in standard diesel engines and is thus distinct from the vegetable and waste oils used to fuel *converted* diesel engines. Biodiesel can be used alone, or blended with petrodiesel in any proportions. Biodiesel blends can also be used as heating oil.

The National Biodiesel Board (USA) also has a technical definition of "biodiesel" as a mono-alkyl ester.

Blends

Biodiesel sample

Blends of biodiesel and conventional hydrocarbon-based diesel are products most commonly distributed for use in the retail diesel fuel marketplace. Much of the world uses a system known as the "B" factor to state the amount of biodiesel in any fuel mix:

- 100% biodiesel is referred to as B100

- 20% biodiesel, 80% petrodiesel is labeled B20

- 5% biodiesel, 95% petrodiesel is labeled B5

- 2% biodiesel, 98% petrodiesel is labeled B2

Blends of 20% biodiesel and lower can be used in diesel equipment with no, or only minor modifications, although certain manufacturers do not extend warranty coverage if equipment is damaged by these blends. The B6 to B20 blends are covered by the ASTM D7467 specification. Biodiesel can also be used in its pure form (B100), but may require certain engine modifications to avoid maintenance and performance problems. Blending B100 with petroleum diesel may be accomplished by:

- Mixing in tanks at manufacturing point prior to delivery to tanker truck
- Splash mixing in the tanker truck (adding specific percentages of biodiesel and petroleum diesel)
- In-line mixing, two components arrive at tanker truck simultaneously.
- Metered pump mixing, petroleum diesel and biodiesel meters are set to X total volume, transfer pump pulls from two points and mix is complete on leaving pump.

Applications

Biodiesel can be used in pure form (B100) or may be blended with petroleum diesel at any concentration in most injection pump diesel engines. New extreme high-pressure (29,000 psi) common rail engines have strict factory limits of B5 or B20, depending on manufacturer. Biodiesel has different solvent properties than petrodiesel, and will degrade natural rubber gaskets and hoses in vehicles (mostly vehicles manufactured before 1992), although these tend to wear out naturally and most likely will have already been replaced with FKM, which is nonreactive to biodiesel. Biodiesel has been known to break down deposits of residue in the fuel lines where petrodiesel has been used. As a result, fuel filters may become clogged with particulates if a quick transition to pure biodiesel is made. Therefore, it is recommended to change the fuel filters on engines and heaters shortly after first switching to a biodiesel blend.

Distribution

Since the passage of the Energy Policy Act of 2005, biodiesel use has been increasing in the United States. In the UK, the Renewable Transport Fuel Obligation obliges suppliers to include 5% renewable fuel in all transport fuel sold in the UK by 2010. For road diesel, this effectively means 5% biodiesel (B5).

Vehicular Use and Manufacturer Acceptance

In 2005, Chrysler (then part of DaimlerChrysler) released the Jeep Liberty CRD diesels from the factory into the American market with 5% biodiesel blends, indicating at least partial acceptance of biodiesel as an acceptable diesel fuel additive. In 2007, Daimler-Chrysler indicated its intention to increase warranty coverage to 20% biodiesel blends if biofuel quality in the United States can be standardized.

The Volkswagen Group has released a statement indicating that several of its vehicles are compatible with B5 and B100 made from rape seed oil and compatible with the EN 14214 standard. The use of the specified biodiesel type in its cars will not void any warranty.

Mercedes Benz does not allow diesel fuels containing greater than 5% biodiesel (B5) due to concerns about "production shortcomings". Any damages caused by the use of such non-approved fuels will not be covered by the Mercedes-Benz Limited Warranty.

Starting in 2004, the city of Halifax, Nova Scotia decided to update its bus system to allow the fleet of city buses to run entirely on a fish-oil based biodiesel. This caused the city some initial mechanical issues, but after several years of refining, the entire fleet had successfully been converted.

In 2007, McDonalds of UK announced it would start producing biodiesel from the waste oil byproduct of its restaurants. This fuel would be used to run its fleet.

The 2014 Chevy Cruze Clean Turbo Diesel, direct from the factory, will be rated for up to B20 (blend of 20% biodiesel / 80% regular diesel) biodiesel compatibility

Railway Usage

Biodiesel locomotive and its external fuel tank at Mount Washington Cog Railway

British train operating company Virgin Trains claimed to have run the UK's first "bio-diesel train", which was converted to run on 80% petrodiesel and 20% biodiesel.

The Royal Train on 15 September 2007 completed its first ever journey run on 100% biodiesel fuel supplied by Green Fuels Ltd. His Royal Highness, The Prince of Wales, and Green Fuels managing director, James Hygate, were the first passengers on a train fueled entirely by biodiesel fuel. Since 2007, the Royal Train has operated successfully on B100 (100% biodiesel).

Similarly, a state-owned short-line railroad in eastern Washington ran a test of a 25% biodiesel / 75% petrodiesel blend during the summer of 2008, purchasing fuel from a biodiesel producer sited along the railroad tracks. The train will be powered by biodiesel made in part from canola grown in agricultural regions through which the short line runs.

Also in 2007, Disneyland began running the park trains on B98 (98% biodiesel). The program was discontinued in 2008 due to storage issues, but in January 2009, it was announced that the park would then be running all trains on biodiesel manufactured from its own used cooking oils. This is a change from running the trains on soy-based biodiesel.

In 2007, the historic Mt. Washington Cog Railway added the first biodiesel locomotive to its all-steam locomotive fleet. The fleet has climbed up the western slopes of Mount Washington in New Hampshire since 1868 with a peak vertical climb of 37.4 degrees.

On 8 July 2014, the then Indian Railway Minister D.V. Sadananda Gowda announced in Railway Budget that 5% bio-diesel will be used in Indian Railways' Diesel Engines.

Aircraft Use

A test flight has been performed by a Czech jet aircraft completely powered on bio-diesel. Other recent jet flights using biofuel, however, have been using other types of renewable fuels.

On November 7, 2011 United Airlines flew the world's first commercial aviation flight on a microbially derived biofuel using Solajet™, Solazyme's algae-derived renewable jet fuel. The Eco-skies Boeing 737-800 plane was fueled with 40 percent Solajet and 60 percent petroleum-derived jet fuel. The commercial Eco-skies flight 1403 departed from Houston's IAH airport at 10:30 and landed at Chicago's ORD airport at 13:03.

As a Heating Oil

Biodiesel can also be used as a heating fuel in domestic and commercial boilers, a mix of heating oil and biofuel which is standardized and taxed slightly differently from diesel fuel used for transportation. Bioheat® fuel is a proprietary blend of biodiesel and traditional heating oil. Bioheat® is a registered trademark of the National Biodiesel Board [NBB] and the National Oilheat Research Alliance [NORA] in the U.S., and Columbia Fuels in Canada). Heating biodiesel is available in various blends. ASTM 396 recognizes blends of up to 5 percent biodiesel as equivalent to pure petroleum heating oil. Blends of higher levels of up to 20% biofuel are used by many consumers. Research is underway to determine whether such blends affect performance.

Older furnaces may contain rubber parts that would be affected by biodiesel's solvent properties, but can otherwise burn biodiesel without any conversion required. Care must be taken, however, given that varnishes left behind by petrodiesel will be released and can clog pipes- fuel filtering and prompt filter replacement is required. Another approach is to start using biodiesel as a blend, and decreasing the petroleum proportion over time can allow the varnishes to come off more gradually and be less likely to clog. Thanks to its strong solvent properties, however, the furnace is cleaned out and generally becomes more efficient. A technical research paper describes laboratory research and field trials project using pure biodiesel and biodiesel blends as a heating fuel in oil-fired boilers. During the Biodiesel Expo 2006 in the UK, Andrew J. Robertson presented his biodiesel heating oil research from his technical paper and suggested B20 biodiesel could reduce UK household CO_2 emissions by 1.5 million tons per year.

A law passed under Massachusetts Governor Deval Patrick requires all home heating diesel in that state to be 2% biofuel by July 1, 2010, and 5% biofuel by 2013. New York City has passed a similar law.

Cleaning Oil Spills

With 80-90% of oil spill costs invested in shoreline cleanup, there is a search for more efficient and cost-effective methods to extract oil spills from the shorelines. Biodiesel has displayed its capacity to significantly dissolve crude oil, depending on the source of the fatty acids. In a laboratory setting, oiled sediments that simulated polluted shorelines were sprayed with a single coat of biodiesel and exposed to simulated tides. Biodiesel is an effective solvent to oil due to its methyl ester component, which considerably lowers the viscosity of the crude oil. Additionally, it has a higher buoyancy than crude oil, which later aids in its removal. As a result, 80% of oil was removed from cobble and fine sand, 50% in coarse sand, and 30% in gravel. Once the oil is liberated from the shoreline, the oil-biodiesel mixture is manually removed from the water surface with skimmers. Any remaining mixture is easily broken down due to the high biodegradability of biodiesel, and the increased surface area exposure of the mixture.

Biodiesel in Generators

Biodiesel is also used in rental generators

In 2001, UC Riverside installed a 6-megawatt backup power system that is entirely fueled by biodiesel. Backup diesel-fueled generators allow companies to avoid damaging blackouts of critical operations at the expense of high pollution and emission rates. By using B100, these generators were able to essentially eliminate the byproducts that result in smog, ozone, and sulfur emissions. The use of these generators in residential areas around schools, hospitals, and the general public result in substantial reductions in poisonous carbon monoxide and particulate matter.

Historical Background

Rudolf Diesel

Transesterification of a vegetable oil was conducted as early as 1853 by Patrick Duffy, four decades before the first diesel engine became functional. Rudolf Diesel's prime model, a single 10 ft (3.0 m) iron cylinder with a flywheel at its base, ran on its own power for the first time in Augsburg, Germany, on 10 August 1893 running on nothing but peanut oil. In remembrance of this event, 10 August has been declared "International Biodiesel Day".

It is often reported that Diesel designed his engine to run on peanut oil, but this is not the case. Diesel stated in his published papers, "at the Paris Exhibition in 1900 (Exposition Universelle) there was shown by the Otto Company a small Diesel engine, which, at the request of the French government ran on arachide (earth-nut or pea-nut) oil, and worked so smoothly that only a few people were aware of it. The engine was constructed for using mineral oil, and was then worked on vegetable oil without any alterations being made. The French Government at the time thought of testing the applicability to power production of the Arachide, or earth-nut, which grows in considerable quantities in their African colonies, and can easily be cultivated there." Diesel himself later conducted related tests and appeared supportive of the idea. In a 1912 speech Diesel said, "the use of vegetable oils for engine fuels may seem insignificant today but such oils may become, in the course of time, as important as petroleum and the coal-tar products of the present time."

Despite the widespread use of petroleum-derived diesel fuels, interest in vegetable oils as fuels for internal combustion engines was reported in several countries during the 1920s and 1930s and later during World War II. Belgium, France, Italy, the United Kingdom, Portugal, Germany, Brazil, Argentina, Japan and China were reported to have tested and used vegetable oils as diesel fuels during this time. Some operational problems were reported due to the high viscosity of vegetable oils compared to petroleum diesel fuel, which results in poor atomization of the fuel in the fuel spray and often leads to deposits and coking of the injectors, combustion chamber and valves. Attempts to overcome these problems included heating of the vegetable oil, blending it with petroleum-derived diesel fuel or ethanol, pyrolysis and cracking of the oils.

On 31 August 1937, G. Chavanne of the University of Brussels (Belgium) was granted a patent for a "Procedure for the transformation of vegetable oils for their uses as fuels" (fr. *"Procédé de Transformation d'Huiles Végétales en Vue de Leur Utilisation comme Carburants"*) Belgian Patent 422,877. This patent described the alcoholysis (often referred to as transesterification) of vegetable oils using ethanol (and mentions methanol) in order to separate the fatty acids from the glycerol by replacing the glycerol with short linear alcohols. This appears to be the first account of the production of what is known as "biodiesel" today.

More recently, in 1977, Brazilian scientist Expedito Parente invented and submitted for patent, the first industrial process for the production of biodiesel. This process is classified as biodiesel by international norms, conferring a "standardized identity and quality. No other proposed biofuel has been validated by the motor industry." As of 2010, Parente's company Tecbio is working with Boeing and NASA to certify bioquerosene (bio-kerosene), another product produced and patented by the Brazilian scientist.

Research into the use of transesterified sunflower oil, and refining it to diesel fuel standards, was initiated in South Africa in 1979. By 1983, the process for producing fuel-quality, engine-tested biodiesel was completed and published internationally. An Austrian company, Gaskoks, obtained the technology from the South African Agricultural Engineers; the company erected the first biodiesel pilot plant in November 1987, and the first industrial-scale plant in April 1989 (with a capacity of 30,000 tons of rapeseed per annum).

Throughout the 1990s, plants were opened in many European countries, including the Czech Republic, Germany and Sweden. France launched local production of biodiesel fuel (referred to as *diester*) from rapeseed oil, which is mixed into regular diesel fuel at a level of 5%, and into the diesel fuel used by some captive fleets (e.g. public transportation) at a level of 30%. Renault, Peugeot and other manufacturers have certified truck engines for use with up to that level of partial biodiesel; experiments with 50% biodiesel are underway. During the same period, nations in other parts of the world also saw local production of biodiesel starting up: by 1998, the Austrian Biofuels Institute had identified 21 countries with commercial biodiesel projects. 100% biodiesel is now available at many normal service stations across Europe.

Properties

Biodiesel has promising lubricating properties and cetane ratings compared to low sulfur diesel fuels. Depending on the engine, this might include high pressure injection pumps, pump injectors (also called *unit injectors*) and fuel injectors.

The calorific value of biodiesel is about 37.27 MJ/kg. This is 9% lower than regular

Number 2 petrodiesel. Variations in biodiesel energy density is more dependent on the feedstock used than the production process. Still, these variations are less than for petrodiesel. It has been claimed biodiesel gives better lubricity and more complete combustion thus increasing the engine energy output and partially compensating for the higher energy density of petrodiesel.

Older diesel Mercedes are popular for running on biodiesel.

The color of biodiesel ranges from golden to dark brown, depending on the production method. It is slightly miscible with water, has a high boiling point and low vapor pressure. *The flash point of biodiesel (>130 °C, >266 °F) is significantly higher than that of petroleum diesel (64 °C, 147 °F) or gasoline (−45 °C, -52 °F). Biodiesel has a density of ~ 0.88 g/cm³, higher than petrodiesel (~ 0.85 g/cm³).

Biodiesel contains virtually no sulfur, and it is often used as an additive to Ultra-Low Sulfur Diesel (ULSD) fuel to aid with lubrication, as the sulfur compounds in petrodiesel provide much of the lubricity.

Fuel Efficiency

The power output of biodiesel depends on its blend, quality, and load conditions under which the fuel is burnt. The thermal efficiency for example of B100 as compared to B20 will vary due to the differing energy content of the various blends. Thermal efficiency of a fuel is based in part on fuel characteristics such as: viscosity, specific density, and flash point; these characteristics will change as the blends as well as the quality of biodiesel varies. The American Society for Testing and Materials has set standards in order to judge the quality of a given fuel sample.

Regarding brake thermal efficiency one study found that B40 was superior to traditional counterpart at higher compression ratios (this higher brake thermal efficiency was recorded at compression ratios of 21:1). It was noted that, as the compression ratios increased, the efficiency of all fuel types - as well as blends being tested - increased; though it was found that a blend of B40 was the most economical at a compression ratio of 21:1 over all other blends. The study implied that this increase in efficiency was due to fuel density, viscosity, and heating values of the fuels.

Combustion

Fuel systems on the modern diesel engine were not designed to accommodate biodiesel, while many heavy duty engines are able to run with biodiesel blends e.g. B20. Traditional direct injection fuel systems operate at roughly 3,000 psi at the injector tip while the modern common rail fuel system operates upwards of 30,000 PSI at the injector tip. Components are designed to operate at a great temperature range, from below freezing to over 1,000 degrees Fahrenheit. Diesel fuel is expected to burn efficiently and produce as few emissions as possible. As emission standards are being introduced to diesel engines the need to control harmful emissions is being designed into the parameters of diesel engine fuel systems. The traditional inline injection system is more forgiving to poorer quality fuels as opposed to the common rail fuel system. The higher pressures and tighter tolerances of the common rail system allows for greater control over atomization and injection timing. This control of atomization as well as combustion allows for greater efficiency of modern diesel engines as well as greater control over emissions. Components within a diesel fuel system interact with the fuel in a way to ensure efficient operation of the fuel system and so the engine. If an out-of-specification fuel is introduced to a system that has specific parameters of operation, then the integrity of the overall fuel system may be compromised. Some of these parameters such as spray pattern and atomization are directly related to injection timing.

One study found that during atomization biodiesel and its blends produced droplets were greater in diameter than the droplets produced by traditional petrodiesel. The smaller droplets were attributed to the lower viscosity and surface tension of traditional petrol. It was found that droplets at the periphery of the spray pattern were larger in diameter than the droplets at the center this was attributed to the faster pressure drop at the edge of the spray pattern; there was a proportional relationship between the droplet size and the distance from the injector tip. It was found that B100 had the greatest spray penetration, this was attributed to the greater density of B100. Having a greater droplet size can lead to; inefficiencies in the combustion, increased emissions, and decreased horse power. In another study it was found that there is a short injection delay when injecting biodiesel. This injection delay was attributed to the greater viscosity of Biodiesel. It was noted that the higher viscosity and the greater cetane rating of biodiesel over traditional petrodiesel lead to poor atomization, as well as mixture penetration with air during the ignition delay period. Another study noted that this ignition delay may aid in a decrease of NOx emission.

Emissions

Emissions are inherent to the combustion of diesel fuels that are regulated by the U.S. Environmental Protection Agency (E.P.A.). As these emissions are a byproduct of the combustion process, in order to ensure E.P.A. compliance a fuel system must be capable of controlling the combustion of fuels as well as the mitigation of emissions. There are a number of new technologies being phased in to control the production of diesel emissions. The exhaust gas recirculation system, E.G.R., and the diesel particulate filter, D.P.F., are both designed to mitigate the production of harmful emissions.

A study performed by the Chonbuk National University concluded that a B30 biodiesel blend reduced carbon monoxide emissions by approximately 83% and particulate matter emissions by roughly 33%. NOx emissions, however, were found to increase without the application of an E.G.R. system. The study also concluded that, with E.G.R, a B20 biodiesel blend considerably reduced the emissions of the engine. Additionally, analysis by the California Air Resources Board found that biodiesel had the lowest carbon emissions of the fuels tested, those being ultra-low-sulfur diesel, gasoline, corn-based ethanol, compressed natural gas, and five types of biodiesel from varying feedstocks. Their conclusions also showed great variance in carbon emissions of biodiesel based on the feedstock used. Of soy, tallow, canola, corn, and used cooking oil, soy showed the highest carbon emissions, while used cooking oil produced the lowest.

While studying the effect of biodiesel on a D.P.F. it was found that though the presence of sodium and potassium carbonates aided in the catalytic conversion of ash, as the diesel particulates are catalyzed, they may congregate inside the D.P.F. and so interfere with the clearances of the filter. This may cause the filter to clog and interfere with the regeneration process. In a study on the impact of E.G.R. rates with blends of jathropa biodiesel it was shown that there was a decrease in fuel efficiency and torque output due to the use of biodiesel on a diesel engine designed with an E.G.R. system. It was found that CO and CO2 emissions increased with an increase in exhaust gas recirculation but NOx levels decreased. The opacity level of the jathropa blends was in an acceptable range, where traditional diesel was out of acceptable standards. It was shown that a decrease in Nox emissions could be obtained with an E.G.R. system. This study showed an advantage over traditional diesel within a certain operating range of the E.G.R. system. Currently blended biodiesel fuels (B5 and B20) are being used in many heavy-duty vehicles especially transit buses in US cities. Characterization of exhaust emissions showed significant emission reductions compared to regular diesel.

Material Compatibility

- Plastics: High-density polyethylene (HDPE) is compatible but polyvinyl chloride (PVC) is slowly degraded. Polystyrene is dissolved on contact with biodiesel.

- Metals: Biodiesel (like methanol) has an effect on copper-based materials (e.g. brass), and it also affects zinc, tin, lead, and cast iron. Stainless steels (316 and 304) and aluminum are unaffected.

- Rubber: Biodiesel also affects types of natural rubbers found in some older engine components. Studies have also found that fluorinated elastomers (FKM) cured with peroxide and base-metal oxides can be degraded when biodiesel loses its stability caused by oxidation. Commonly used synthetic rubbers FKM- GBL-S and FKM-GF-S found in modern vehicles were found to handle biodiesel in all conditions.

Technical Standards

Biodiesel has a number of standards for its quality including European standard EN 14214, ASTM International D6751, and others.

Low Temperature Gelling

When biodiesel is cooled below a certain point, some of the molecules aggregate and form crystals. The fuel starts to appear cloudy once the crystals become larger than one quarter of the wavelengths of visible light - this is the cloud point (CP). As the fuel is cooled further these crystals become larger. The lowest temperature at which fuel can pass through a 45 micrometre filter is the cold filter plugging point (CFPP). As biodiesel is cooled further it will gel and then solidify. Within Europe, there are differences in the CFPP requirements between countries. This is reflected in the different national standards of those countries. The temperature at which pure (B100) biodiesel starts to gel varies significantly and depends upon the mix of esters and therefore the feedstock oil used to produce the biodiesel. For example, biodiesel produced from low erucic acid varieties of canola seed (RME) starts to gel at approximately −10 °C (14 °F). Biodiesel produced from beef tallow and palm oil tends to gel at around 16 °C (61 °F) and 13 °C (55 °F) respectively. There are a number of commercially available additives that will significantly lower the pour point and cold filter plugging point of pure biodiesel. Winter operation is also possible by blending biodiesel with other fuel oils including #2 low sulfur diesel fuel and #1 diesel / kerosene.

Another approach to facilitate the use of biodiesel in cold conditions is by employing a second fuel tank for biodiesel in addition to the standard diesel fuel tank. The second fuel tank can be insulated and a heating coil using engine coolant is run through the tank. The fuel tanks can be switched over when the fuel is sufficiently warm. A similar method can be used to operate diesel vehicles using straight vegetable oil.

Contamination by Water

Biodiesel may contain small but problematic quantities of water. Although it is only slightly miscible with water it is hygroscopic. One of the reasons biodiesel can absorb water is the persistence of mono and diglycerides left over from an incomplete reaction. These molecules can act as an emulsifier, allowing water to mix with the biodiesel. In addition, there may be water that is residual to processing or resulting from storage tank condensation. The presence of water is a problem because:

- Water reduces the heat of fuel combustion, causing smoke, harder starting, and reduced power.

- Water causes corrosion of fuel system components (pumps, fuel lines, etc.)

- Microbes in water cause the paper-element filters in the system to rot and fail, causing failure of the fuel pump due to ingestion of large particles.

- Water freezes to form ice crystals that provide sites for nucleation, accelerating gelling of the fuel.

- Water causes pitting in pistons.

Previously, the amount of water contaminating biodiesel has been difficult to measure by taking samples, since water and oil separate. However, it is now possible to measure the water content using water-in-oil sensors.

Water contamination is also a potential problem when using certain chemical catalysts involved in the production process, substantially reducing catalytic efficiency of base (high pH) catalysts such as potassium hydroxide. However, the super-critical methanol production methodology, whereby the transesterification process of oil feedstock and methanol is effectuated under high temperature and pressure, has been shown to be largely unaffected by the presence of water contamination during the production phase.

Availability and Prices

In some countries biodiesel is less expensive than conventional diesel

Global biodiesel production reached 3.8 million tons in 2005. Approximately 85% of biodiesel production came from the European Union.

In 2007, in the United States, average retail (at the pump) prices, including federal and state fuel taxes, of B2/B5 were lower than petroleum diesel by about 12 cents, and B20 blends were the same as petrodiesel. However, as part of a dramatic shift in diesel pricing, by July 2009, the US DOE was reporting average costs of B20 15 cents per gallon higher than petroleum diesel ($2.69/gal vs. $2.54/gal). B99 and B100 generally cost more than petrodiesel except where local governments provide a tax incentive or subsidy.

Production

Biodiesel is commonly produced by the transesterification of the vegetable oil or animal fat feedstock. There are several methods for carrying out this transesterification reaction including the common batch process, supercritical processes, ultrasonic methods, and even microwave methods.

Chemically, transesterified biodiesel comprises a mix of mono-alkyl esters of long chain fatty acids. The most common form uses methanol (converted to sodium methoxide) to produce methyl esters (commonly referred to as Fatty Acid Methyl Ester - FAME) as it is the cheapest alcohol available, though ethanol can be used to produce an ethyl ester (commonly referred to as Fatty Acid Ethyl Ester - FAEE) biodiesel and higher alcohols such as isopropanol and butanol have also been used. Using alcohols of higher molecular weights improves the cold flow properties of the resulting ester, at the cost of a less efficient transesterification reaction. A lipid transesterification production process is used to convert the base oil to the desired esters. Any free fatty acids (FFAs) in the base oil are either converted to soap and removed from the process, or they are esterified (yielding more biodiesel) using an acidic catalyst. After this processing, unlike straight vegetable oil, biodiesel has combustion properties very similar to those of petroleum diesel, and can replace it in most current uses.

The methanol used in most biodiesel production processes is made using fossil fuel inputs. However, there are sources of renewable methanol made using carbon dioxide or biomass as feedstock, making their production processes free of fossil fuels.

A by-product of the transesterification process is the production of glycerol. For every 1 tonne of biodiesel that is manufactured, 100 kg of glycerol are produced. Originally, there was a valuable market for the glycerol, which assisted the economics of the process as a whole. However, with the increase in global biodiesel production, the market price for this crude glycerol (containing 20% water and catalyst residues) has crashed. Research is being conducted globally to use this glycerol as a chemical building block. One initiative in the UK is The Glycerol Challenge.

Usually this crude glycerol has to be purified, typically by performing vacuum distillation. This is rather energy intensive. The refined glycerol (98%+ purity) can then be utilised directly, or converted into other products. The following announcements were made in 2007: A joint venture of Ashland Inc. and Cargill announced plans to make propylene glycol in Europe from glycerol and Dow Chemical announced similar plans for North America. Dow also plans to build a plant in China to make epichlorhydrin from glycerol. Epichlorhydrin is a raw material for epoxy resins.

Production Levels

In 2007, biodiesel production capacity was growing rapidly, with an average annual growth rate from 2002-06 of over 40%. For the year 2006, the latest for which actual production figures could be obtained, total world biodiesel production was about 5-6 million tonnes, with 4.9 million tonnes processed in Europe (of which 2.7 million tonnes was from Germany) and most of the rest from the USA. In 2008 production in Europe alone had risen to 7.8 million tonnes. In July 2009, a duty was added to American imported biodiesel in the European Union in order to balance the competition from European, especially German producers. The capacity for 2008 in Europe totalled 16 million tonnes. This compares with a total demand for diesel in the US and Europe of approximately 490 million tonnes (147 billion gallons). Total world production of vegetable oil for all purposes in 2005/06 was about 110 million tonnes, with about 34 million tonnes each of palm oil and soybean oil.

US biodiesel production in 2011 brought the industry to a new milestone. Under the EPA Renewable Fuel Standard, targets have been implemented for the biodiesel production plants in order to monitor and document production levels in comparison to total demand. According to the year-end data released by the EPA, biodiesel production in 2011 reached more than 1 billion gallons. This production number far exceeded the 800 million gallon target set by the EPA. The projected production for 2020 is nearly 12 billion gallons.

Biodiesel Feedstocks

A variety of oils can be used to produce biodiesel. These include:

- Virgin oil feedstock – rapeseed and soybean oils are most commonly used, soybean oil accounting for about half of U.S. production. It also can be obtained from Pongamia, field pennycress and jatropha and other crops such as mustard, jojoba, flax, sunflower, palm oil, coconut and hemp;

- Waste vegetable oil (WVO);

- Animal fats including tallow, lard, yellow grease, chicken fat, and the by-products of the production of Omega-3 fatty acids from fish oil.

- Algae, which can be grown using waste materials such as sewage and without displacing land currently used for food production.

- Oil from halophytes such as Salicornia bigelovii, which can be grown using saltwater in coastal areas where conventional crops cannot be grown, with yields equal to the yields of soybeans and other oilseeds grown using freshwater irrigation

- Sewage Sludge - The sewage-to-biofuel field is attracting interest from major companies like Waste Management and startups like InfoSpi, which are betting that renewable sewage biodiesel can become competitive with petroleum diesel on price.

Many advocates suggest that waste vegetable oil is the best source of oil to produce biodiesel, but since the available supply is drastically less than the amount of petroleum-based fuel that is burned for transportation and home heating in the world, this local solution could not scale to the current rate of consumption.

Animal fats are a by-product of meat production and cooking. Although it would not be efficient to raise animals (or catch fish) simply for their fat, use of the by-product adds value to the livestock industry (hogs, cattle, poultry). Today, multi-feedstock biodiesel facilities are producing high quality animal-fat based biodiesel. Currently, a 5-million dollar plant is being built in the USA, with the intent of producing 11.4 million litres (3 million gallons) biodiesel from some of the estimated 1 billion kg (2.2 billion pounds) of chicken fat produced annually at the local Tyson poultry plant. Similarly, some small-scale biodiesel factories use waste fish oil as feedstock. An EU-funded project (ENERFISH) suggests that at a Vietnamese plant to produce biodiesel from catfish (basa, also known as pangasius), an output of 13 tons/day of biodiesel can be produced from 81 tons of fish waste (in turn resulting from 130 tons of fish). This project utilises the biodiesel to fuel a CHP unit in the fish processing plant, mainly to power the fish freezing plant.

Quantity of Feedstocks Required

Current worldwide production of vegetable oil and animal fat is not sufficient to replace liquid fossil fuel use. Furthermore, some object to the vast amount of farming and the resulting fertilization, pesticide use, and land use conversion that would be needed to produce the additional vegetable oil. The estimated transportation diesel fuel and home heating oil used in the United States is about 160 million tons (350 billion pounds) according to the Energy Information Administration, US Department of Energy. In the United States, estimated production of vegetable oil for all uses is about 11 million tons (24 billion pounds) and estimated production of animal fat is 5.3 million tonnes (12 billion pounds).

If the entire arable land area of the USA (470 million acres, or 1.9 million square kilometers) were devoted to biodiesel production from soy, this would just about provide the 160 million tonnes required (assuming an optimistic 98 US gal/acre of biodiesel). This land area could in principle be reduced significantly using algae, if the obstacles can be overcome. The US DOE estimates that if algae fuel replaced all the petroleum fuel in the United States, it would require 15,000 square miles (39,000 square kilometers), which is a few thousand square miles larger than Maryland, or 30% greater than the area of Belgium, assuming a yield of 140 tonnes/hectare (15,000 US gal/acre). Given a more realistic yield of 36 tonnes/hectare (3834 US gal/acre) the area required is about 152,000 square kilometers, or roughly equal to that of the state of Georgia or of England and Wales. The advantages of algae are that it can be grown on non-arable land such as deserts or in marine environments, and the potential oil yields are much higher than from plants.

Yield

Feedstock yield efficiency per unit area affects the feasibility of ramping up production to the huge industrial levels required to power a significant percentage of vehicles.

Some typical yields		
Crop	**Yield**	
	L/ha	**US gal/acre**
Palm oil	4752	508
Coconut	2151	230
Cyperus esculentus	1628	174
Rapeseed	954	102
Soy (Indiana)	554-922	59.2-98.6
Chinese tallow	907	97
Peanut	842	90
Sunflower	767	82
Hemp	242	26

1. "Biofuels: some numbers". *Grist.org. Retrieved 2010-03-15.*

2. Makareviciene et al., "Opportunities for the use of chufa sedge in biodiesel production", Industrial Crops and Products, 50 (2013) p. 635, table 2.

3. Klass, Donald, "Biomass for Renewable Energy, Fuels, and Chemicals", page 341. Academic Press, 1998.

4. Kitani, Osamu, "Volume V: Energy and Biomass Engineering, CIGR Handbook of Agricultural Engineering", Amer Society of Agricultural, 1999.

Algae fuel yields have not yet been accurately determined, but DOE is reported as saying that algae yield 30 times more energy per acre than land crops such as soybeans. Yields of 36 tonnes/hectare are considered practical by Ami Ben-Amotz of the Institute of Oceanography in Haifa, who has been farming Algae commercially for over 20 years.

Jatropha has been cited as a high-yield source of biodiesel but yields are highly dependent on climatic and soil conditions. The estimates at the low end put the yield at about 200 US gal/acre (1.5-2 tonnes per hectare) per crop; in more favorable climates two or more crops per year have been achieved. It is grown in the Philippines, Mali and India, is drought-resistant, and can share space with other cash crops such as coffee, sugar, fruits and vegetables. It is well-suited to semi-arid lands and can contribute to slow down desertification, according to its advocates.

Efficiency and Economic Arguments

Pure biodiesel (B-100) made from soybeans

According to a study by Drs. Van Dyne and Raymer for the Tennessee Valley Authority, the average US farm consumes fuel at the rate of 82 litres per hectare (8.75 US gal/acre) of land to produce one crop. However, average crops of rapeseed produce oil at an average rate of 1,029 L/ha (110 US gal/acre), and high-yield rapeseed fields produce about 1,356 L/ha (145 US gal/acre). The ratio of input to output in these cases is roughly 1:12.5 and 1:16.5. Photosynthesis is known to have an efficiency rate of about 3-6% of total solar radiationand if the entire mass of a crop is utilized for energy production, the overall efficiency of this chain is currently about 1%While this may compare unfavorably to solar cells combined with an electric drive train, biodiesel is less costly to deploy (solar cells cost approximately US$250 per square meter) and transport (electric vehicles require batteries which currently have a much lower energy density than liquid fuels). A 2005 study found that biodiesel production using soybeans required 27% more fossil energy than the biodiesel produced and 118% more energy using sunflowers.

However, these statistics by themselves are not enough to show whether such a change makes economic sense. Additional factors must be taken into account, such as: the fuel equivalent of the energy required for processing, the yield of fuel from raw oil, the return on cultivating food, the effect biodiesel will have on food prices and the relative cost of biodiesel versus petrodiesel, water pollution from farm run-off, soil depletion, and the externalized costs of political and military interference in oil-producing countries intended to control the price of petrodiesel.

The debate over the energy balance of biodiesel is ongoing. Transitioning fully to biofuels could require immense tracts of land if traditional food crops are used (although non food crops can be utilized). The problem would be especially severe for nations with large economies, since energy consumption scales with economic output.

If using only traditional food plants, most such nations do not have sufficient arable land to produce biofuel for the nation's vehicles. Nations with smaller economies (hence less energy consumption) and more arable land may be in better situations, although many regions cannot afford to divert land away from food production.

For third world countries, biodiesel sources that use marginal land could make more sense; e.g., pongam oiltree nuts grown along roads or jatropha grown along rail lines.

In tropical regions, such as Malaysia and Indonesia, plants that produce palm oil are being planted at a rapid pace to supply growing biodiesel demand in Europe and other markets. Scientists have shown that the removal of rainforest for palm plantations is not ecologically sound since the expansion of oil palm plantations poses a threat to natural rainforest and biodiversity.

It has been estimated in Germany that palm oil biodiesel has less than one third of the production costs of rapeseed biodiesel.The direct source of the energy content of biodiesel is solar energy captured by plants during photosynthesis. Regarding the positive energy balance of biodiesel:

When straw was left in the field, biodiesel production was strongly energy positive, yielding 1 GJ biodiesel for every 0.561 GJ of energy input (a yield/cost ratio of 1.78).

When straw was burned as fuel and oilseed rapemeal was used as a fertilizer, the yield/cost ratio for biodiesel production was even better (3.71). In other words, for every unit of energy input to produce biodiesel, the output was 3.71 units (the difference of 2.71 units would be from solar energy).

Economic Impact

Multiple economic studies have been performed regarding the economic impact of biodiesel production. One study, commissioned by the National Biodiesel Board, reported the 2011 production of biodiesel supported 39,027 jobs and more than $2.1 billion in household income. The growth in biodiesel also helps significantly increase GDP. In 2011, biodiesel created more than $3 billion in GDP. Judging by the continued growth in the Renewable Fuel Standard and the extension of the biodiesel tax incentive, the number of jobs can increase to 50,725, $2.7 billion in income, and reaching $5 billion in GDP by 2012 and 2013.

Energy Security

One of the main drivers for adoption of biodiesel is energy security. This means that a nation's dependence on oil is reduced, and substituted with use of locally available sources, such as coal, gas, or renewable sources. Thus a country can benefit from adoption of biofuels, without a reduction in greenhouse gas emissions. While the total energy balance is debated, it is clear that the dependence on oil is reduced. One example is the energy used to manufacture fertilizers, which could come from a variety of sources other than petroleum. The US National Renewable Energy Laboratory (NREL) states that energy security is the number one driving force behind the US biofuels programme, and a White House "Energy Security for the 21st Century" paper makes it clear that energy security is a major reason for promoting biodiesel. The former EU commission president, Jose Manuel Barroso, speaking at a recent EU biofuels conference, stressed that properly managed biofuels have the potential to reinforce the EU's security of supply through diversification of energy sources.

Global Biofuel Policies

Many countries around the world are involved in the growing use and production of biofuels, such as biodiesel, as an alternative energy source to fossil fuels and oil. To foster the biofuel industry, governments have implemented legislations and laws as incentives to reduce oil dependency and to increase the use of renewable energies. Many countries have their own independent policies regarding the taxation and rebate of biodiesel use, import, and production.

Canada

It was required by the Canadian Environmental Protection Act Bill C-33 that by the year 2010, gasoline contained 5% renewable content and that by 2013, diesel and heating oil contained 2% renewable content. The EcoENERGY for Biofuels Program subsidized the production of biodiesel, among other biofuels, via an incentive rate of CAN$0.20 per liter from 2008 to 2010. A decrease of $0.04 will be applied every year following, until the incentive rate reaches $0.06 in 2016. Individual provinces also have specific legislative measures in regards to biofuel use and production.

United States

The Volumetric Ethanol Excise Tax Credit (VEETC) was the main source of financial support for biofuels, but was scheduled to expire in 2010. Through this act, biodiesel production guaranteed a tax credit of US$1 per gallon produced from virgin oils, and $0.50 per gallon made from recycled oils. Currently soybean oil is being used to produce soybean biodiesel for many commercial purposes such as blending fuel for transportation sectors.

European Union

The European Union is the greatest producer of biodiesel, with France and Germany being the top producers. To increase the use of biodiesel, there exist policies requiring the blending of biodiesel into fuels, including penalties if those rates are not reached. In France, the goal was to reach 10% integration but plans for that stopped in 2010. As an incentive for the European Union countries to continue the production of the biofuel, there are tax rebates for specific quotas of biofuel produced. In Germany, the minimum percentage of biodiesel in transport diesel is set at 7% so called "B7".

Environmental Effects

The surge of interest in biodiesels has highlighted a number of environmental effects associated with its use. These potentially include reductions in greenhouse gas emissions,deforestation, pollution and the rate of biodegradation.

According to the EPA's Renewable Fuel Standards Program Regulatory Impact Analysis, released in February 2010, biodiesel from soy oil results, on average, in a 57% reduction in greenhouse gases compared to petroleum diesel, and biodiesel produced from waste grease results in an 86% reduction.

However, environmental organizations, for example, Rainforest Rescueand Greenpeace,criticize the cultivation of plants used for biodiesel production, e.g., oil palms, soybeans and sugar cane. They say the deforestation of rainforests exacerbates climate change and that sensitive ecosystems are destroyed to clear land for oil palm, soybean and sugar cane plantations. Moreover, that biofuels contribute to world hunger, seeing as arable land is no longer used for growing foods. The Environmental Protection Agency(EPA) published data in January 2012, showing that biofuels made from palm oil won't count towards the nation's renewable fuels mandate as they are not climate-friendly.Environmentalists welcome the conclusion because the growth of oil palm plantations has driven tropical deforestation, for example, in Indonesia and Malaysia.

Food, Land and Water vs. Fuel

In some poor countries the rising price of vegetable oil is causing problems.Some propose that fuel only be made from non-edible vegetable oils such as camelina, jatropha or seashore mallowwhich can thrive on marginal agricultural land where many trees and crops will not grow, or would produce only low yields.

Others argue that the problem is more fundamental. Farmers may switch from producing food crops to producing biofuel crops to make more money, even if the new crops are not edible.The law of supply and demand predicts that if fewer farmers are producing food the price of food will rise. It may take some time, as farmers can take some time to change which things they are growing, but increasing demand for first generation biofuels is likely to result in price increases for many kinds of food. Some have pointed out that there are poor farmers and poor countries who are making more money because of the higher price of vegetable oil.

Biodiesel from sea algae would not necessarily displace terrestrial land currently used for food production and new algaculture jobs could be created.

Current Research

There is ongoing research into finding more suitable crops and improving oil yield.

Other sources are possible including human fecal matter, with Ghana building its first "fecal sludge-fed biodiesel plant."Using the current yields, vast amounts of land and fresh water would be needed to produce enough oil to completely replace fossil fuel usage. It would require twice the land area of the US to be devoted to soybean production, or two-thirds to be devoted to rapeseed production, to meet current US heating and transportation needs.

Specially bred mustard varieties can produce reasonably high oil yields and are very useful in crop rotation with cereals, and have the added benefit that the meal leftover after the oil has been pressed out can act as an effective and biodegradable pesticide.

The NFESC, with Santa Barbara-based Biodiesel Industries is working to develop biodiesel technologies for the US navy and military, one of the largest diesel fuel users in the world.

A group of Spanish developers working for a company called Ecofasa announced a new biofuel made from trash. The fuel is created from general urban waste which is treated by bacteria to produce fatty acids, which can be used to make biodiesel.

Another approach that does not require the use of chemical for the production involves the use of genetically modified microbes.

Algal Biodiesel

From 1978 to 1996, the U.S. NREL experimented with using algae as a biodiesel source in the "Aquatic Species Program". A self-published article by Michael Briggs, at the UNH Biodiesel Group, offers estimates for the realistic replacement of all vehicular fuel with biodiesel by utilizing algae that have a natural oil content greater than 50%, which Briggs suggests can be grown on algae ponds at wastewater treatment plants. This oil-rich algae can then be extracted from the system and processed into biodiesel, with the dried remainder further reprocessed to create ethanol.

The production of algae to harvest oil for biodiesel has not yet been undertaken on a commercial scale, but feasibility studies have been conducted to arrive at the above yield estimate. In addition to its projected high yield, algaculture — unlike crop-based biofuels — does not entail a decrease in food production, since it requires neither farmland nor fresh water. Many companies are pursuing algae bio-reactors for various purposes, including scaling up biodiesel production to commercial levels.

Prof. Rodrigo E. Teixeira from the University of Alabama in Huntsville demonstrated the extraction of biodiesel lipids from wet algae using a simple and economical reaction in ionic liquids.

Pongamia

Millettia pinnata, also known as the Pongam Oiltree or Pongamia, is a leguminous, oilseed-bearing tree that has been identified as a candidate for non-edible vegetable oil production.

Pongamia plantations for biodiesel production have a two-fold environmental benefit. The trees both store carbon and produce fuel oil. Pongamia grows on marginal land not fit for food crops and does not require nitrate fertilizers. The oil producing tree has the highest yield of oil producing plant (approximately 40% by weight of the seed is oil) while growing in malnourished soils with high levels of salt. It is becoming a main focus in a number of biodiesel research organizations.The main advantages of Pongamia are a higher recovery and quality of oil than other crops and no direct competition with food crops. However, growth on marginal land can lead to lower oil yields which could cause competition with food crops for better soil.

Jatropha

Jatropha Biodiesel from DRDO, India.

Several groups in various sectors are conducting research on Jatropha curcas, a poisonous shrub-like tree that produces seeds considered by many to be a viable source of biodiesel feedstock oil.Much of this research focuses on improving the overall per acre oil yield of Jatropha through advancements in genetics, soil science, and horticultural practices.

SG Biofuels, a San Diego-based Jatropha developer, has used molecular breeding and biotechnology to produce elite hybrid seeds of Jatropha that show significant yield improvements over first generation varieties.SG Biofuels also claims that additional benefits have arisen from such strains, including improved flowering synchronicity, higher resistance to pests and disease, and increased cold weather tolerance.

Plant Research International, a department of the Wageningen University and Research Centre in the Netherlands, maintains an ongoing Jatropha Evaluation Project (JEP) that examines the feasibility of large scale Jatropha cultivation through field and laboratory experiments.

The Center for Sustainable Energy Farming (CfSEF) is a Los Angeles-based non-profit research organization dedicated to Jatropha research in the areas of plant science, agronomy, and horticulture. Successful exploration of these disciplines is projected to increase Jatropha farm production yields by 200-300% in the next ten years.

Fungi

A group at the Russian Academy of Sciences in Moscow published a paper in September 2008, stating that they had isolated large amounts of lipids from single-celled fungi and turned it into biodiesel in an economically efficient manner. More research on this fungal species; Cunninghamella japonica, and others, is likely to appear in the near future.

The recent discovery of a variant of the fungus Gliocladium roseum points toward the production of so-called myco-diesel from cellulose. This organism was recently discovered in the rainforests of northern Patagonia and has the unique capability of converting cellulose into medium length hydrocarbons typically found in diesel fuel.

Biodiesel from Used Coffee Grounds

Researchers at the University of Nevada, Reno, have successfully produced biodiesel from oil derived from used coffee grounds. Their analysis of the used grounds showed a 10% to 15% oil content (by weight). Once the oil was extracted, it underwent conventional processing into biodiesel. It is estimated that finished biodiesel could be produced for about one US dollar per gallon. Further, it was reported that "the technique is not difficult" and that "there is so much coffee around that several hundred million gallons of biodiesel could potentially be made annually." However, even if all the coffee grounds in the world were used to make fuel, the amount produced would be less than 1 percent of the diesel used in the United States annually. "It won't solve the world's energy problem," Dr. Misra said of his work.

Exotic Sources

Recently, alligator fat was identified as a source to produce biodiesel. Every year, about 15 million pounds of alligator fat are disposed of in landfills as a waste byproduct of the alligator meat and skin industry. Studies have shown that biodiesel produced from alligator fat is similar in composition to biodiesel created from soybeans, and is cheaper to refine since it is primarily a waste product.

Biodiesel to Hydrogen-cell Power

A microreactor has been developed to convert biodiesel into hydrogen steam to power fuel cells.

Steam reforming, also known as fossil fuel reforming is a process which produces hydrogen gas from hydrocarbon fuels, most notably biodiesel due to its efficiency. A **microreactor**, or reformer, is the processing device in which water vapour reacts with the liquid fuel under high temperature and pressure. Under temperatures ranging from $700 - 1100$ °C, a nickel-based catalyst enables the production of carbon monoxide and hydrogen:

$$Hydrocarbon + H_2O \rightleftharpoons CO + 3\,H_2 \text{ (Highly endothermic)}$$

Furthermore, a higher yield of hydrogen gas can be harnessed by further oxidizing carbon monoxide to produce more hydrogen and carbon dioxide:

$$CO + H_2O \rightarrow CO_2 + H_2 \text{ (Mildly exothermic)}$$

Hydrogen Fuel Cells Background Information

Fuel cells operate similar to a battery in that electricity is harnessed from chemical reactions. The difference in fuel cells when compared to batteries is their ability to be powered by the constant flow of hydrogen found in the atmosphere. Furthermore, they produce only water as a by-product, and are virtually silent. The downside of hydrogen powered fuel cells is the high cost and dangers of storing highly combustible hydrogen under pressure.

One way new processors can overcome the dangers of transporting hydrogen is to produce it as necessary. The microreactors can be joined to create a system that heats the hydrocarbon under high pressure to generate hydrogen gas and carbon dioxide, a process called steam reforming. This produces up to 160 gallons of hydrogen/minute and gives the potential of powering hydrogen refueling stations, or even an on-board hydrogen fuel source for hydrogen cell vehicles.Implementation into cars would allow energy-rich fuels, such as biodiesel, to be transferred to kinetic energy while avoiding combustion and pollutant byproducts. The hand-sized square piece of metal contains microscopic channels with catalytic sites, which continuously convert biodiesel, and even its glycerol byproduct, to hydrogen.

Concerns

Engine Wear

Lubricity of fuel plays an important role in wear that occurs in an engine. An engine relies on its fuel to provide lubricity for the metal components that are constantly in contact with each other.Biodiesel is a much better lubricant compared with petroleum diesel due to the presence of esters. Tests have shown that the addition of a small amount of biodiesel to diesel can significantly increase the lubricity of the fuel in short term.However, over a longer period of time (2–4 years), studies show that biodiesel loses its lubricity.This could be because of enhanced corrosion over time due to oxidation of the unsaturated molecules or increased water content in biodiesel from moisture absorption.

Fuel Viscosity

One of the main concerns regarding biodiesel is its viscosity. The viscosity of diesel is 2.5–3.2 cSt at 40 °C and the viscosity of biodiesel made from soybean oil is between 4.2 and 4.6 cStThe viscosity of diesel must be high enough to provide sufficient lubrication for the engine parts but low enough to flow at operational temperature. High viscosity can plug the fuel filter and injection system in engines.Vegetable oil is composed of lipids with long chains of hydrocarbons, to reduce its viscosity the lipids are broken down into smaller molecules of esters. This is done by converting vegetable oil and animal fats into alkyl esters using transesterification to reduce their viscosityNevertheless, biodiesel viscosity remains higher than that of diesel, and the engine may not be able to use the fuel at low temperatures due to the slow flow through the fuel filter.

Engine Performance

Biodiesel has higher brake-specific fuel consumption compared to diesel, which means more biodiesel fuel consumption is required for the same torque. However, B20 biodiesel blend has been found to provide maximum increase in thermal efficiency, lowest brake-specific energy consumption, and lower harmful emissions. The engine performance depends on the properties of the fuel, as well as on combustion, injector pressure and many other factors.Since there are various blends of biodiesel, that may account for the contradicting reports in regards engine performance.

Pellet Fuel

Wood pellets

Pellet fuels (or pellets) are biofuels made from compressed organic matter or biomass. Pellets can be made from any one of five general categories of biomass: industrial waste and co-products, food waste, agricultural residues, energy crops, and virgin lumber. Wood pellets are the most common type of pellet fuel and are generally made from compacted sawdust and related industrial wastes from the milling of lumber, manufacture of wood products and furniture, and construction. Other industrial waste sources include empty fruit bunches, palm kernel shells, coconut shells, and tree tops and branches discarded during logging operations. So-called "black pellets" are made of biomass, refined to resemble hard coal and were developed to be used in existing coal-fired power plants. Pellets are categorized by their heating value, moisture and ash content, and dimensions. They can be used as fuels for power generation, commercial or residential heating, and cooking. Pellets are extremely dense and can be produced with a low moisture content (below 10%) that allows them to be burned with a very high combustion efficiency.

Fuels for heating
• Heating oil
• Wood pellet
• Kerosene
• Propane
• Natural gas
• Wood
• Coal

Further, their regular geometry and small size allow automatic feeding with very fine calibration. They can be fed to a burner by auger feeding or by pneumatic conveying. Their high density also permits compact storage and transport over long distance. They can be conveniently blown from a tanker to a storage bunker or silo on a customer's premises.

A broad range of pellet stoves, central heating furnaces, and other heating appliances have been developed and marketed since the mid-1980s. In 1997 fully automatic wood pellet boilers with similar comfort level as oil and gas boilers became available in Austria. With the surge in the price of fossil fuels since 2005, the demand for pellet heating has increased in Europe and North America, and a sizable industry is emerging. According to the International Energy Agency Task 40, wood pellet production has more than doubled between 2006 and 2010 to over 14 million tons. In a 2012 report, the Biomass Energy Resource Center says that it expects wood pellet production in North America to double again in the next five years.

Production

Pellets are produced by compressing the wood material which has first passed through a hammer mill to provide a uniform dough-like mass. This mass is fed to a press, where it is squeezed through a die having holes of the size required (normally 6 mm diameter, sometimes 8 mm or larger). The high pressure of the press causes the temperature of the wood to increase greatly, and the lignin plasticizes slightly, forming a natural "glue" that holds the pellet together as it cools.

Pellets can be made from grass and other non-woody forms of biomass that do not contain lignin: distiller's dried grains (a brewing industry byproduct) can be added to provide the necessary durability. A 2005 news story from Cornell University News suggested that grass pellet production was more advanced in Europe than North America. It suggested the benefits of grass as a feedstock included its short growing time (70 days), and ease of cultivation and processing. The story quoted Jerry Cherney, an agriculture professor at the school, stating that grasses produce 96% of the heat of wood and that "any mixture of grasses can be used, cut in mid- to late summer, left in the field to leach out minerals, then baled and pelleted. Drying of the hay is not required for pelleting, making the cost of processing less than with wood pelleting." In 2012, the Department of Agriculture of Nova Scotia announced as a demonstration project conversion of an oil-fired boiler to grass pellets at a research facility.

Rice-husk fuel-pellets are made by compacting rice-husk obtained as by-product of rice-growing from the fields. It also has similar characteristics to the wood-pellets and more environment-friendly, as the raw material is a waste-product. The energy content is about 4-4.2 kcal/kg and moisture content is typically less than 10%. The size of pellets is generally kept to be about 6mm diameter and 25mm length in the form of a cylinder; though larger cylinder or briquette forms are not uncommon. It is much cheaper than similar energy-pellets and can be compacted/manufactured from the husk at the farm itself, using cheap machinery. They generally are more environment-friendly as compared to wood-pellets. In the regions of the world where wheat is the predominant food-crop, wheat husk can also be compacted to produce energy-pellets, with characteristics similar to rice-husk pellets.

Pellet truck being filled at a plant in Germany.

A report by CORRIM (Consortium On Research on Renewable Industrial Material) for the Life-Cycle Inventory of Wood Pellet Manufacturing and Utilization estimates the energy required to dry, pelletize and transport pellets is less than 11% of the energy content of the pellets if using pre-dried industrial wood waste. If the pellets are made directly from forest material, it takes up to 18% of the energy to dry the wood and additional 8% for transportation and manufacturing energy. An environmental impact assessment of exported wood pellets by the Department of Chemical and Mineral Engineering, University of Bologna, Italy and the Clean Energy Research Centre, at the University of British Columbia, published in 2009, concluded that the energy consumed to ship Canadian wood pellets from Vancouver to Stockholm (15,500 km via the Panama Canal), is about 14% of the total energy content of the wood pellets.

Pellet Standards

Pellets conforming to the norms commonly used in Europe (DIN 51731 or Ö-Norm M-7135) have less than 10% water content, are uniform in density (higher than 1 ton per cubic meter, thus it sinks in water)(bulk density about 0.6-0.7 ton per cubic meter), have good structural strength, and low dust and ash content. Because the wood fibres are broken down by the hammer mill, there is virtually no difference in the finished pellets between different wood types. Pellets can be made from nearly any wood variety, provided the pellet press is equipped with good instrumentation, the differences in feed material can be compensated for in the press regulation.. In Europe, the main production areas are located in south Scandinavia, Finland, Central Europe, Austria, and the Baltic countries.

Pellets conforming to the European standards norms which contain recycled wood or outside contaminants are considered Class B pellets. Recycled materials such as particle board, treated or painted wood, melamine resin-coated panels and the like are particularly unsuitable for use in pellets, since they may produce noxious emissions and uncontrolled variations in the burning characteristics of the pellets.

Standards used in the United States are different, developed by the Pellet Fuels Institute and, as in Europe, are not mandatory. Still, many manufacturers comply, as warranties of US-manufactured or imported combustion equipment may not cover damage by pellets non-conformant with regulations. Prices for US pellets surged during the fossil fuel price inflation of 2007–2008, but later dropped markedly and are generally lower on a per-BTU basis than most fossil fuels, excluding coal.

Regulatory agencies in Europe and North America are in the process of tightening the emissions standards for all forms of wood heat, including wood pellets and pellet stoves. These standards will become mandatory, with independently certified testing to ensure compliance. In the United States, the new rules initiated in 2009 have completed the EPA regulatory review process, with final new Standards of Performance for New Residential Wood Heaters rules issued for comment on June 24, 2014. The American Lumber Standard Committee will be the independent certification agency for the new pellet standards.

Pellet Stove Operation

There are three general types of pellet heating appliances, free standing pellet stoves, pellet stove inserts and pellet boilers. *Pellet stoves* "look like traditional wood stoves but operate more like a modern furnace. [Fuel, wood or other biomass pellets, is stored in a storage bin called a hopper. The hopper can be located on the top of the appliance, the side of it or remotely.] A mechanical auger [automatically feeds] the pellets into a burn pot, where they are incinerated at such a high temperature that they create no vent-clogging creosote and very little ash or emissions... "Heat-exchange tubes": Send air heated by fire into room... "Convection fan": Circulates air through heat-exchange tubes and into room... The biggest difference between a pellet stove and ... a woodstove, is that, inside, the pellet stove is a high-tech device with a circuit board, a thermostat, and fans—all of which work together to [regulate temperature and] heat your space efficiently."

A *pellet stove insert* is a stove that is inserted into an existing masonry or prefabricated wood fireplace.

Pellet boilers are standalone central heating and hot water systems designed to replace traditional fossil fuel systems in residential, commercial and institutional applications. Automatic or *auto-pellet boilers* include silos for bulk storage of pellets, a fuel delivery system that moves the fuel from the silo to the hopper, a logic controller to regulate temperature across multiple heating zones and an automated ash removal system for long-term automated operations.

Pellet baskets allow a person to heat their home using pellets in existing stoves or fireplaces.

Energy Output and Efficiency

Wood-pellet heater

The energy content of wood pellets is approximately 4.7 – 5.2 MWh/tonne (~7450 BTU/lb).

High-efficiency wood pellet stoves and boilers have been developed in recent years, typically offering combustion efficiencies of over 85%. The newest generation of wood pellet boilers can work in condensing mode and therefore achieve 12% higher efficiency values. Wood pellet boilers have limited control over the rate and presence of combustion compared to liquid or gaseous-fired systems; however, for this reason they are better suited for hydronic heating systems due to the hydronic system's greater ability to store heat. Pellet burners capable of being retrofitted to oil-burning boilers are also available.

Air Pollution Emissions

Emissions such as NO_x, SO_x and volatile organic compounds from pellet burning equipment are in general very low in comparison to other forms of combustion heating. A recognized problem is the emission of fine particulate matter to the air, especially in urban areas that have a high concentration of pellet heating systems or coal or oil heating systems in close proximity. This $PM_{2.5}$ emissions of older pellet stoves and boilers can be problematic in close quarters, especially in comparison to natural gas (or renewable biogas), though on large installations electrostatic precipitators, cyclonic separators, or baghouse particle filters can control particulates when properly maintained and operated.

Global Warming

There is uncertainty to what degree making heat or electricity by burning wood pellets contributes to global climate change, as well as how the impact on climate compares to the impact of using competing sources of heat. Factors in the uncertainty include the wood source, carbon dioxide emissions from production and transport as well as from final combustion, and what time scale is appropriate for the consideration.

A report by the Manomet Center for Conservation Sciences, "Biomass Sustainability and Carbon Policy Study" issued in June 2010 for the Massachusetts Department of Energy Resources, concludes that burning biomass such as wood pellets or wood chips releases a large amount of CO2 into the air, creating a "carbon debt" that is not retired for 20–25 years and after which there is a net benefit. In June 2011 the department was preparing to file its final regulation, expecting to significantly tighten controls on the use of biomass for energy, including wood pellets. Biomass energy proponents have disputed the Manomet report's conclusions, and scientists have pointed out oversights in the report, suggesting that climate impacts are worse than reported.

Until ca. 2008 it was commonly assumed, even in scientific papers, that biomass energy (including from wood pellets) is carbon neutral, largely because regrowth of vegetation was believed to recapture and store the carbon that is emitted to the air. Then, scientific papers studying the climate implications of biomass began to appear which refuted the simplistic assumption of its carbon neutrality. According to the Biomass Energy Resource Center, the assumption of carbon neutrality "has shifted to a recognition that the carbon implications of biomass depend on how the fuel is harvested, from what forest types, what kinds of forest management are applied, and how biomass is used over time and across the landscape."

In 2011 twelve prominent U.S. environmental organizations adopted policy setting a high bar for government incentives of biomass energy, including wood pellets. It states in part that, "[b]iomass sources and facilities qualifying for (government) incentives must result in lower life-cycle, cumulative and net GHG and ocean acidifying emissions, within 20 years and also over the longer term, than the energy sources they replace or compete with."

Sustainability

The wood products industry is concerned that if large-scale use of wood energy is instituted, the supply of raw materials for construction and manufacturing will be significantly curtailed.

Cost

Due to the rapid increase in popularity since 2005, pellet availability and cost may be an issue. This is an important consideration when buying a pellet stove, furnace, pellet baskets or other devices known in the industry as Bradley Burners. However, current pellet production is increasing and there are plans to bring several new pellet mills online in the US in 2008–2009.

The cost of the pellets can be affected by the building cycle leading to fluctuations in the supply of sawdust and offcuts.

Per the New Hampshire Office of Energy and Planning release on Fuel Prices updated on 5 Oct 2015, the cost of #2 Fuel Oil delivered can be compared to the cost of Bulk Delivered Wood Fuel Pellets using their BTU equivalent: 1 ton pellets = 118.97 gallon of #2 Fuel Oil. This assumes that one ton of pellets produces 16,500,000 BTU and one gallon of #2 Fuel Oil produces 138,690 BTU. Thus if #2 Fuel Oil delivered costs $1.90/Gal, the breakeven price for pellets is $238.00/Ton delivered.

Usage by Region

Europe

EU Pellet Use (ton)	
Country	**2013**
UK	4 540 000
Italy	3 300 000
Denmark	2 500 000
Netherlands	2 000 000
Sweden	1 650 000
Germany	1 600 000
Belgium	1 320 000

Usage across Europe varies due to government regulations. In the Netherlands, Belgium, and the UK, pellets are used mainly in large-scale power plants. In Denmark and Sweden, pellets are used in large-scale power plants, medium-scale district heating systems, and small-scale residential heat. In Germany, Austria, Italy, and France, pellets are used mostly for small-scale residential and industrial heat.

The UK has initiated a grant scheme called the Renewable Heat Incentive (RHI) allowing non-domestic and domestic wood pellet boiler installations to receive payments over a period of between 7–20 years It is the first such scheme in the world and aims to increase the amount of renewable energy generated in the UK, in line with EU commitments. Scotland and Northern Ireland have separate but similar schemes. From Spring 2015, any biomass owners whether domestic or commercial must buy their fuels from BSL (Biomass Suppliers List) approved suppliers in order to receive RHI payments.

Pellets are widely used in Sweden, the main pellet producer in Europe, mainly as an alternative to oil-fired central heating. In Austria, the leading market for pellet central heating furnaces (relative to its population), it is estimated that 2/3 of all new domestic heating furnaces are pellet burners. In Italy, a large market for automatically fed pellet stoves has developed. Italy's main usage for pellets is small - scale private residential and industrial boilers for heating.

In 2014 in Germany the overall wood pellet consumption per year comprised 2,2 mln tones. These pellets are consumed predominantly by residential small scale heating sector. The co-firing plants which use pellet sector for energy production are not widespread in the country. The largest amount of wood pellets is certified with DINplus and these are the pellets of the highest quality. As a rule, the pellets of lower quality are exported.

New Zealand

The total sales of wood pellets in New Zealand was 3–5,000 tonnes in 2003. Recent construction of new wood pellet plants has given a huge increase in production capacity.

United States

Some companies import European-made boilers. As of 2009, about 800,000 Americans were using wood pellets for heat. It is estimated that 2.33 million tons of wood pellets will be used for heat in the US in 2013.

Other Uses

Horse Bedding

When small amounts of water are added to wood pellets, they expand and revert to sawdust. This makes them suitable to use as a horse bedding. The ease of storage and transportation are additional benefits over traditional bedding. However, some species of wood, including walnut, can be toxic to horses and should never be used for bedding.

In Thailand, rice husk pellets are being produced for animal bedding. They have a high absorption rate which makes them ideal for the purpose.

Absorbents

Wood pellets are also used to absorb contaminated water when drilling oil or gas wells.

Ethanol Fuel

The Saab 9-3 SportCombi BioPower was the second E85 flexifuel model introduced by Saab in the Swedish market.

Ethanol fuel is ethyl alcohol, the same type of alcohol found in alcoholic beverages, used as fuel. It is most often used as a motor fuel, mainly as a biofuel additive for gasoline. The first production car running entirely on ethanol was the Fiat 147, introduced in 1978 in Brazil by Fiat. Nowadays, cars are able to run using 100% ethanol fuel or a mix of ethanol and gasoline (aka flex-fuel). It is commonly made from biomass such as corn or sugarcane. World ethanol production for transport fuel tripled between 2000 and 2007 from 17 billion to more than 52 billion liters. From 2007 to 2008, the share of ethanol in global gasoline type fuel use increased from 3.7% to 5.4%. In 2011 worldwide ethanol fuel production reached 22.36 billion U.S. liquid gallons (bg) (84.6 billion liters), with the United States as the top producer with 13.9 bg (52.6 billion liters), accounting for 62.2% of global production, followed by Brazil with 5.6 bg (21.1 billion liters). Ethanol fuel has a "gasoline gallon equivalency" (GGE) value of 1.5 US gallons (5.7 L), which means 1.5 gallons of ethanol produces the energy of one gallon of gasoline.

Ethanol fuel is widely used in Brazil and in the United States, and together both countries were responsible for 87.1% of the world's ethanol fuel production in 2011. Most cars on the road today in the U.S. can run on blends of up to 10% ethanol, and ethanol represented 10% of the U.S. gasoline fuel supply derived from domestic sources in 2011. Since 1976 the Brazilian government has made it mandatory to blend ethanol with gasoline, and since 2007 the legal blend is around 25% ethanol and 75% gasoline (E25). By December 2011 Brazil had a fleet of 14.8 million flex-fuel automobiles and light trucks and 1.5 million flex-fuel motorcycles that regularly use neat ethanol fuel (known as E100).

Bioethanol is a form of quasi-renewable energy that can be produced from agricultural feedstocks. It can be made from very common crops such as hemp, sugarcane, potato, cassava and corn. There has been considerable debate about how useful bioethanol is in replacing gasoline. Concerns about its production and use relate to increased food prices due to the large amount of arable land required for crops, as well as the energy and pollution balance of the whole cycle of ethanol production, especially from corn. Recent developments with cellulosic ethanol production and commercialization may allay some of these concerns.

Cellulosic ethanol offers promise because cellulose fibers, a major and universal component in plant cells walls, can be used to produce ethanol. According to the International Energy Agency, cellulosic ethanol could allow ethanol fuels to play a much bigger role in the future.

Chemistry

Structure of ethanol molecule. All bonds are single bonds

During ethanol fermentation, glucose and other sugars in the corn (or sugarcane or other crops) are converted into ethanol and carbon dioxide.

$$C_6H_{12}O_6 \rightarrow 2\ C_2H_5OH + 2\ CO_2 + \text{heat}$$

Ethanol fermentation is not 100% selective with other side products such as acetic acid, glycols and many other products produced. They are mostly removed during ethanol purification. Fermentation takes place in an aqueous solution. The resulting solution has an ethanol content of around 15%. Ethanol is subsequently isolated and purified by a combination of adsorption and distillation.

During combustion, ethanol reacts with oxygen to produce carbon dioxide, water, and heat:

$$C_2H_5OH + 3\ O_2 \rightarrow 2\ CO_2 + 3\ H_2O + \text{heat}$$

Starch and cellulose molecules are strings of glucose molecules. It is also possible to generate ethanol out of cellulosic materials. That, however, requires a pretreatment that splits the cellulose into glycose molecules and other sugars that subsequently can be fermented. The resulting product is called cellulosic ethanol, indicating its source.

Ethanol may also be produced industrially from ethylene by hydration of the double bond in the presence of catalysts and high temperature.

$$C_2H_4 + H_2O \rightarrow C_2H_5OH$$

By far the largest fraction of the global ethanol production, however, is produced by fermentation.

Sources

Sugar cane harvest

Ethanol is a quasi-renewable energy source because while the energy is partially generated by using a resource, sunlight, which cannot be depleted, the harvesting process requires vast amounts of energy that typically comes from non-renewable sources. Creation of ethanol starts with photosynthesis causing a feedstock, such as sugar cane or a grain such as maize (corn), to grow. These feedstocks are processed into ethanol.

Cornfield in South Africa

About 5% of the ethanol produced in the world in 2003 was actually a petroleum product. It is made by the catalytic hydration of ethylene with sulfuric acid as the catalyst. It can also be obtained via ethylene or acetylene, from calcium carbide, coal, oil gas, and other sources. Two million tons of petroleum-derived ethanol are produced annually. The principal suppliers are plants in the United States, Europe, and South Africa. Petroleum derived ethanol (synthetic ethanol) is chemically identical to bio-ethanol and can be differentiated only by radiocarbon dating.

Bio-ethanol is usually obtained from the conversion of carbon-based feedstock. Agricultural feedstocks are considered renewable because they get energy from the sun using photosynthesis, provided that all minerals required for growth (such as nitrogen and phosphorus) are returned to the land. Ethanol can be produced from a variety of feedstocks such as sugar cane, bagasse, miscanthus, sugar beet, sorghum, grain, switchgrass, barley, hemp, kenaf, potatoes, sweet potatoes, cassava, sunflower, fruit, molasses, corn, stover, grain, wheat, straw, cotton, other biomass, as well as many types of cellulose waste and harvesting, whichever has the best well-to-wheel assessment.

Switchgrass

An alternative process to produce bio-ethanol from algae is being developed by the company Algenol. Rather than grow algae and then harvest and ferment it, the algae grow in sunlight and produce ethanol directly, which is removed without killing the algae. It is claimed the process can produce 6,000 US gallons per acre (56,000 litres per ha) per year compared with 400 US gallons per acre (3,750 l/ha) for corn production.

Currently, the first generation processes for the production of ethanol from corn use only a small part of the corn plant: the corn kernels are taken from the corn plant and only the starch, which represents about 50% of the dry kernel mass, is transformed into ethanol. Two types of second generation processes are under development. The first type uses enzymes and yeast fermentation to convert the plant cellulose into ethanol while the second type uses pyrolysis to convert the whole plant to either a liquid bio-oil or a syngas. Second generation processes can also be used with plants such as grasses, wood or agricultural waste material such as straw.

Production

The basic steps for large-scale production of ethanol are: microbial (yeast) fermentation of sugars, distillation, dehydration, and denaturing (optional). Prior to fermentation, some crops require saccharification or hydrolysis of carbohydrates such as cellulose and starch into sugars. Saccharification of cellulose is called cellulolysis. Enzymes are used to convert starch into sugar.

Fermentation

Ethanol is produced by microbial fermentation of the sugar. Microbial fermentation currently only works directly with sugars. Two major components of plants, starch and cellulose, are both made of sugars—and can, in principle, be converted to sugars for fermentation. Currently, only the sugar (e.g., sugar cane) and starch (e.g., corn) portions can be economically converted. There is much activity in the area of cellulosic ethanol, where the cellulose part of a plant is broken down to sugars and subsequently converted to ethanol.

Distillation

Ethanol plant in West Burlington, Iowa

For the ethanol to be usable as a fuel, the majority of the water must be removed. Most of the water is removed by distillation, but the purity is limited to 95–96% due to the formation of a low-boiling water-ethanol azeotrope with maximum (95.6% m/m (96.5% v/v) ethanol and 4.4% m/m (3.5% v/v) water). This mixture is called hydrous ethanol and can be used as a fuel alone, but unlike anhydrous ethanol, hydrous ethanol is not miscible in all ratios with gasoline, so the water fraction is typically removed in further treatment to burn in combination with gasoline in gasoline engines.

Ethanol plant in Sertãozinho, Brazil.

Dehydration

There are basically three dehydration processes to remove the water from an azeotropic ethanol/water mixture. The first process, used in many early fuel ethanol plants, is called azeotropic distillation and consists of adding benzene or cyclohexane to the mixture. When these components are added to the mixture, it forms a heterogeneous azeotropic mixture in vapor–liquid-liquid equilibrium, which when distilled produces anhydrous ethanol in the column bottom, and a vapor mixture of water, ethanol, and cyclohexane/benzene.

When condensed, this becomes a two-phase liquid mixture. The heavier phase, poor in the entrainer (benzene or cyclohexane), is stripped of the entrainer and recycled to the feed—while the lighter phase, with condensate from the stripping, is recycled to the second column. Another early method, called extractive distillation, consists of adding a ternary component that increases ethanol's relative volatility. When the ternary mixture is distilled, it produces anhydrous ethanol on the top stream of the column.

With increasing attention being paid to saving energy, many methods have been proposed that avoid distillation altogether for dehydration. Of these methods, a third method has emerged and has been adopted by the majority of modern ethanol plants. This new process uses molecular sieves to remove water from fuel ethanol. In this process, ethanol vapor under pressure passes through a bed of molecular sieve beads. The bead's pores are sized to allow absorption of water while excluding ethanol. After a period of time, the bed is regenerated under vacuum or in the flow of inert atmosphere (e.g. N_2) to remove the absorbed water. Two beds are often used so that one is available to absorb water while the other is being regenerated. This dehydration technology can account for energy saving of 3,000 btus/gallon (840 kJ/L) compared to earlier azeotropic distillation.

Post-production Water Issues

Ethanol is hygroscopic, meaning it absorbs water vapor directly from the atmosphere. Because absorbed water dilutes the fuel value of the ethanol and may cause phase separation of ethanol-gasoline blends (which causes engine stall), containers of ethanol fuels must be kept tightly sealed. This high miscibility with water means that ethanol cannot be efficiently shipped through modern pipelines, like liquid hydrocarbons, over long distances. Mechanics also have seen increased cases of damage to small engines, in particular, the carburetor, attributable to the increased water retention by ethanol in fuel.

The fraction of water that an ethanol-gasoline fuel can contain without phase separation increases with the percentage of ethanol. This shows, for example, that E30 can have up to about 2% water. If there is more than about 71% ethanol, the remainder can be any proportion of water or gasoline and phase separation does not occur. The fuel mileage declines with increased water content. The increased solubility of water with higher ethanol content permits E30 and hydrated ethanol to be put in the same tank since any combination of them always results in a single phase. Somewhat less water is tolerated at lower temperatures. For E10 it is about 0.5% v/v at 70 F and decreases to about 0.23% v/v at −30 F.

Consumer Production Systems

While biodiesel production systems have been marketed to home and business users for many years, commercialized ethanol production systems designed for end-consumer use have lagged in the marketplace. In 2008, two different companies announced home-scale ethanol production systems. The AFS125 Advanced Fuel System from Allard Research and Development is capable of producing both ethanol and biodiesel in one machine, while the E-100 MicroFueler from E-Fuel Corporation is dedicated to ethanol only.

Engines

Ethanol most commonly powers Otto cycle internal combustion engines, most often on cars. However, it may be used to power vehicles using a Diesel cycle such as buses and farm tractors. Ethanol has been tested for use as aviation fuel but it is not commercialised .

Fuel Economy

Ethanol contains approx. 34% less energy per unit volume than gasoline, and therefore in theory, burning pure ethanol in a vehicle reduces miles per US gallon 34%, given the same fuel economy, compared to burning pure gasoline. However, since ethanol has a higher octane rating, the engine can be made more efficient by raising its compression ratio. Using a variable turbocharger, the compression ratio can be optimized for the fuel, making fuel economy almost constant for any blend.

For E10 (10% ethanol and 90% gasoline), the effect is small (~3%) when compared to conventional gasoline, and even smaller (1–2%) when compared to oxygenated and re-formulated blends. For E85 (85% ethanol), the effect becomes significant. E85 produces lower mileage than gasoline, and requires more frequent refueling. Actual performance may vary depending on the vehicle. Based on EPA tests for all 2006 E85 models, the average fuel economy for E85 vehicles resulted 25.56% lower than unleaded gasoline. The EPA-rated mileage of current United States flex-fuel vehicles should be considered when making price comparisons, but E85 is a high performance fuel, with an octane rating of about 94–96, and should be compared to premium.

Cold Start During the Winter

High ethanol blends present a problem to achieve enough vapor pressure for the fuel to evaporate and spark the ignition during cold weather (since ethanol tends to increase fuel enthalpy of vaporization). When vapor pressure is below 45 kPa starting a cold engine becomes difficult. To avoid this problem at temperatures below 11 °C (52 °F), and to reduce ethanol higher emissions during cold weather, both the US and the European markets adopted E85 as the maximum blend to be used in their flexible fuel vehicles, and they are optimized to run at such a blend. At places with harsh cold weather, the ethanol blend in the US has a seasonal reduction to E70 for these very cold regions, though it is still sold as E85. At places where temperatures fall below −12 °C (10 °F) during the winter, it is recommended to install an engine heater system, both for gasoline and E85 vehicles. Sweden has a similar seasonal reduction, but the ethanol content in the blend is reduced to E75 during the winter months.

The Brazilian 2008 Honda Civic flex-fuel has outside direct access to the secondary reservoir gasoline tank in the front right side, the corresponding fuel filler door is shown by the arrow.

Brazilian flex fuel vehicles can operate with ethanol mixtures up to E100, which is hydrous ethanol (with up to 4% water), which causes vapor pressure to drop faster as compared to E85 vehicles. As a result, Brazilian flex vehicles are built with a small secondary gasoline reservoir located near the engine. During a cold start pure gasoline is injected to avoid starting problems at low temperatures. This provision is particularly necessary for users of Brazil's southern and central regions, where temperatures normally drop below 15 °C (59 °F) during the winter. An improved flex engine generation was launched in 2009 that eliminates the need for the secondary gas storage tank. In March 2009 Volkswagen do Brasil launched the Polo E-Flex, the first Brazilian flex fuel model without an auxiliary tank for cold start.

Fuel Mixtures

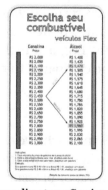

Hydrated ethanol × gasoline type C price table for use in Brazil

In many countries cars are mandated to run on mixtures of ethanol. All Brazilian light-duty vehicles are built to operate for an ethanol blend of up to 25% (E25), and since 1993 a federal law requires mixtures between 22% and 25% ethanol, with 25% required as of mid July 2011. In the United States all light-duty vehicles are built to operate normally with an ethanol blend of 10% (E10). At the end of 2010 over 90 percent of all gasoline sold in the U.S. was blended with ethanol. In January 2011 the U.S. Environmental Protection Agency (EPA) issued a waiver to authorize up to 15% of ethanol blended with gasoline (E15) to be sold only for cars and light pickup trucks with a model year of 2001 or newer. Other countries have adopted their own requirements.

EPA's E15 label required to be displayed in all E15 fuel dispensers in the U.S.

Beginning with the model year 1999, an increasing number of vehicles in the world are manufactured with engines that can run on any fuel from 0% ethanol up to 100% ethanol without modification. Many cars and light trucks (a class containing minivans, SUVs and pickup trucks) are designed to be flexible-fuel vehicles using ethanol blends up to 85% (E85) in North America and Europe, and up to 100% (E100) in Brazil. In older model years, their engine systems contained alcohol sensors in the fuel and/or oxygen sensors in the exhaust that provide input to the engine control computer to adjust the fuel injection to achieve stochiometric (no residual fuel or free oxygen in the exhaust) air-to-fuel ratio for any fuel mix. In newer models, the alcohol sensors have been removed, with the computer using only oxygen and airflow sensor feedback to estimate alcohol content. The engine control computer can also adjust (advance) the ignition timing to achieve a higher output without pre-ignition when it predicts that higher alcohol percentages are present in the fuel being burned. This method is backed up by advanced knock sensors – used in most high performance gasoline engines regardless of whether they are designed to use ethanol or not – that detect pre-ignition and detonation.

Corrosion

Whether ethanol causes unacceptable levels of corrosion in internal combustion engines has been a cause of some debate. In general, E10 blends pose no problems. At higher blends, ethanol's different chemistry from oil derivates makes it corrosive for certain engine parts. These are replaced to create the flex-fuel vehicles that accept up to E85.

Other Engine Configurations

ED95 engines

Since 1989 there have also been ethanol engines based on the diesel principle operating in Sweden. They are used primarily in city buses, but also in distribution trucks and waste collectors. The engines, made by Scania, have a modified compression ratio, and the fuel (known as ED95) used is a mix of 93.6% ethanol and 3.6% ignition improver, and 2.8% denaturants. The ignition improver makes it possible for the fuel to ignite in the diesel combustion cycle. It is then also possible to use the energy efficiency of the diesel principle with ethanol. These engines have been used in the United Kingdom by Reading Transport but the use of bioethanol fuel is now being phased out.

Dual-fuel direct-injection

A 2004 MIT study and an earlier paper published by the Society of Automotive Engineers identified a method to exploit the characteristics of fuel ethanol substantially more efficiently than mixing it with gasoline. The method presents the possibility of leveraging the use of alcohol to achieve definite improvement over the cost-effectiveness of hybrid electric. The improvement consists of using dual-fuel direct-injection of pure alcohol (or the azeotrope or E85) and gasoline, in any ratio up to 100% of either, in a turbocharged, high compression-ratio, small-displacement engine having performance similar to an engine having twice the displacement. Each fuel is carried separately, with a much smaller tank for alcohol. The high-compression (for higher efficiency) engine runs on ordinary gasoline under low-power cruise conditions. Alcohol is directly injected into the cylinders (and the gasoline injection simultaneously reduced) only when necessary to suppress 'knock' such as when significantly accelerating. Direct cylinder injection raises the already high octane rating of ethanol up to an effective 130. The calculated over-all reduction of gasoline use and CO_2 emission is 30%. The consumer cost payback time shows a 4:1 improvement over turbo-diesel and a 5:1 improvement over hybrid. The problems of water absorption into pre-mixed gasoline (causing phase separation), supply issues of multiple mix ratios and cold-weather starting are also avoided.

Increased thermal efficiency

In a 2008 study, complex engine controls and increased exhaust gas recirculation allowed a compression ratio of 19.5 with fuels ranging from neat ethanol to E50. Thermal efficiency up to approximately that for a diesel was achieved. This would result in the fuel economy of a neat ethanol vehicle to be about the same as one burning gasoline.

Fuel cells powered by an ethanol reformer

In June 2016, Nissan announced plans to develop fuel cell vehicles powered by ethanol rather than hydrogen, the fuel of choice by the other car manufacturers that have developed and commercialized fuel cell vehicles, such as the Hyundai Tucson FCEV, Toyota

Mirai, and Honda FCX Clarity. The main advantage of this technical approach is that it would be cheaper and easier to deploy the fueling infrastructure than setting up the one required to deliver hydrogen at high pressures, as each hydrogen fueling station cost US$1 million to US$2 million to build.

Nissan plans to create a technology that uses liquid ethanol fuel as a source to generate hydrogen within the vehicle itself. The technology uses heat to reform ethanol into hydrogen to feed what is known as a solid oxide fuel cell (SOFC). The fuel cell generates electricity to supply power to the electric motor driving the wheels, through a battery that handles peak power demands and stores regenerated energy. The vehicle would include a tank for a blend of water and ethanol, which is fed into an onboard reformer that splits it into pure hydrogen and carbon dioxide. According to Nissan, the liquid fuel could be an ethanol-water blend at a 55:45 ratio. Nissan expects to commercialize its technology by 2020.

Experience by Country

The world's top ethanol fuel producers in 2011 were the United States with 13.9 billion U.S. liquid gallons (bg) (52.60 billion liters) and Brazil with 5.6 bg (21.1 billion liters), accounting together for 87.1% of world production of 22.36 billion US gallons (84.6 billion liters). Strong incentives, coupled with other industry development initiatives, are giving rise to fledgling ethanol industries in countries such as Germany, Spain, France, Sweden, China, Thailand, Canada, Colombia, India, Australia, and some Central American countries.

World rank	Country/Region	2011	2010	2009	2008	2007
	Annual fuel ethanol production by country (2007–2011) Top 10 countries/regional blocks (Millions of U.S. liquid gallons per year)					
1	United States	13,900.00	13,231.00	10,938.00	9,235.00	6,485.00
2	Brazil	5,573.24	6,921.54	6,577.89	6,472.20	5,019.20
3	EU	1,199.31	1,176.88	1,039.52	733.60	570.30
4	China	554.76	541.55	541.55	501.90	486.00
5	Thailand			435.20	89.80	79.20
6	Canada	462.30	356.63	290.59	237.70	211.30
7	India			91.67	66.00	52.80
8	Colombia			83.21	79.30	74.90
9	Australia	87.20	66.04	56.80	26.40	26.40
10	Other			247.27		
	World Total	22,356.09	22,946.87	19,534.99	17,335.20	13,101.70

Environment

Energy Balance

Energy balance		
Country	**Type**	**Energy balance**
United States	Corn ethanol	1.3
Germany	Biodiesel	2.5
Brazil	Sugarcane ethanol	8
United States	Cellulosic ethanol[†]	2–36[††]

† experimental, not in commercial production

†† depending on production method

All biomass goes through at least some of these steps: it needs to be grown, collected, dried, fermented, distilled, and burned. All of these steps require resources and an infrastructure. The total amount of energy input into the process compared to the energy released by burning the resulting ethanol fuel is known as the energy balance (or "energy returned on energy invested"). Figures compiled in a 2007 by National Geographic Magazine point to modest results for corn ethanol produced in the US: one unit of fossil-fuel energy is required to create 1.3 energy units from the resulting ethanol. The energy balance for sugarcane ethanol produced in Brazil is more favorable, with one unit of fossil-fuel energy required to create 8 from the ethanol. Energy balance estimates are not easily produced, thus numerous such reports have been generated that are contradictory. For instance, a separate survey reports that production of ethanol from sugarcane, which requires a tropical climate to grow productively, returns from 8 to 9 units of energy for each unit expended, as compared to corn, which only returns about 1.34 units of fuel energy for each unit of energy expended. A 2006 University of California Berkeley study, after analyzing six separate studies, concluded that producing ethanol from corn uses much less petroleum than producing gasoline.

Carbon dioxide, a greenhouse gas, is emitted during fermentation and combustion. This is canceled out by the greater uptake of carbon dioxide by the plants as they grow to produce the biomass. When compared to gasoline, depending on the production method, ethanol releases less greenhouse gases.

Air Pollution

Compared with conventional unleaded gasoline, ethanol is a particulate-free burning fuel source that combusts with oxygen to form carbon dioxide, carbon monoxide, water and aldehydes. The Clean Air Act requires the addition of oxygenates to reduce carbon monoxide emissions in the United States. The additive MTBE is currently being phased out due to ground water contamination, hence ethanol becomes an attractive alternative additive. Current production methods include air pollution from the manufacturer of macronutrient fertilizers such as ammonia.

A study by atmospheric scientists at Stanford University found that E85 fuel would increase the risk of air pollution deaths relative to gasoline by 9% in Los Angeles, US: a very large, urban, car-based metropolis that is a worst-case scenario. Ozone levels are significantly increased, thereby increasing photochemical smog and aggravating medical problems such as asthma.

Brazil burns significant amounts of ethanol biofuel. Gas chromatograph studies were performed of ambient air in São Paulo, Brazil, and compared to Osaka, Japan, which does not burn ethanol fuel. Atmospheric Formaldehyde was 160% higher in Brazil, and Acetaldehyde was 260% higher.

Manufacture

In 2002, monitoring the process of ethanol production from corn revealed that they released VOCs (volatile organic compounds) at a higher rate than had previously been disclosed. The U.S. Environmental Protection Agency (EPA) subsequently reached settlement with Archer Daniels Midland and Cargill, two of the largest producers of ethanol, to reduce emission of these VOCs. VOCs are produced when fermented corn mash is dried for sale as a supplement for livestock feed. Devices known as thermal oxidizers or catalytic oxidizers can be attached to the plants to burn off the hazardous gases.

Carbon Dioxide

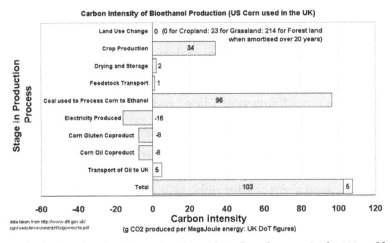

UK government calculation of carbon intensity of corn bioethanol grown in the US and burnt in the UK.

The calculation of exactly how much carbon dioxide is produced in the manufacture of bioethanol is a complex and inexact process, and is highly dependent on the method by which the ethanol is produced and the assumptions made in the calculation. A calculation should include:

- The cost of growing the feedstock

- The cost of transporting the feedstock to the factory

- The cost of processing the feedstock into bioethanol

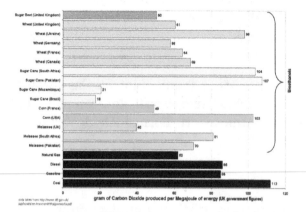

Graph of UK figures for the carbon intensity of bioethanol and fossil fuels.
This graph assumes that all bioethanols are burnt in their country of origin
and that previously existing cropland is used to grow the feedstock.

Such a calculation may or may not consider the following effects:

- The cost of the change in land use of the area where the fuel feedstock is grown.

- The cost of transportation of the bioethanol from the factory to its point of use

- The efficiency of the bioethanol compared with standard gasoline

- The amount of Carbon Dioxide produced at the tail pipe.

- The benefits due to the production of useful bi-products, such as cattle feed or electricity.

The graph on the right shows figures calculated by the UK government for the purposes of the Renewable transport fuel obligation.

The January 2006 Science article from UC Berkeley's ERG, estimated reduction from corn ethanol in GHG to be 13% after reviewing a large number of studies. In a correction to that article released shortly after publication, they reduce the estimated value to 7.4%. A National Geographic Magazine overview article (2007) puts the figures at 22% less CO_2 emissions in production and use for corn ethanol compared to gasoline and a 56% reduction for cane ethanol. Carmaker Ford reports a 70% reduction in CO_2 emissions with bioethanol compared to petrol for one of their flexible-fuel vehicles.

An additional complication is that production requires tilling new soil which produces a one-off release of GHG that it can take decades or centuries of production reductions in GHG emissions to equalize. As an example, converting grass lands to corn production for ethanol takes about a century of annual savings to make up for the GHG released from the initial tilling.

Change in Land Use

Large-scale farming is necessary to produce agricultural alcohol and this requires substantial amounts of cultivated land. University of Minnesota researchers report that if all corn grown in the U.S. were used to make ethanol it would displace 12% of current U.S. gasoline consumption. There are claims that land for ethanol production is acquired through deforestation, while others have observed that areas currently supporting forests are usually not suitable for growing crops. In any case, farming may involve a decline in soil fertility due to reduction of organic matter, a decrease in water availability and quality, an increase in the use of pesticides and fertilizers, and potential dislocation of local communities. New technology enables farmers and processors to increasingly produce the same output using less inputs.

Cellulosic ethanol production is a new approach that may alleviate land use and related concerns. Cellulosic ethanol can be produced from any plant material, potentially doubling yields, in an effort to minimize conflict between food needs vs. fuel needs. Instead of utilizing only the starch by-products from grinding wheat and other crops, cellulosic ethanol production maximizes the use of all plant materials, including gluten. This approach would have a smaller carbon footprint because the amount of energy-intensive fertilisers and fungicides remain the same for higher output of usable material. The technology for producing cellulosic ethanol is currently in the commercialization stage.

Using Biomass for Electricity Instead of Ethanol

Converting biomass to electricity for charging electric vehicles may be a more "climate-friendly" transportation option than using biomass to produce ethanol fuel, according to an analysis published in Science in May 2009 Researchers continue to search for more cost-effective developments in both cellulosic ethanol and advanced vehicle batteries.

Health Costs of Ethanol Emissions

For each billion ethanol-equivalent gallons of fuel produced and combusted in the US, the combined climate-change and health costs are $469 million for gasoline, $472–952 million for corn ethanol depending on biorefinery heat source (natural gas, corn stover, or coal) and technology, but only $123–208 million for cellulosic ethanol depending on feedstock (prairie biomass, Miscanthus, corn stover, or switchgrass).

Efficiency of Common Crops

As ethanol yields improve or different feedstocks are introduced, ethanol production may become more economically feasible in the US. Currently, research on improving ethanol yields from each unit of corn is underway using biotechnology. Also, as long as oil prices remain high, the economical use of other feedstocks, such as cellulose, become viable. By-products such as straw or wood chips can be converted to ethanol. Fast growing species like switchgrass can be grown on land not suitable for other cash crops and yield high levels of ethanol per unit area.

Crop	Annual yield (Liters/hect-are, US gal/acre)	Green-house-gas savings vs. petrol[a]	Comments
Sugar cane	6800–8000 L/ha, 727–870 g/acre	87%–96%	Long-season annual grass. Used as feedstock for most bioethanol produced in Brazil. Newer processing plants burn residues not used for ethanol to generate electricity. Grows only in tropical and subtropical climates.
Miscanthus	7300 L/ha, 780 g/acre	37%–73%	Low-input perennial grass. Ethanol production depends on development of cellulosic technology.
Switchgrass	3100–7600 L/ha, 330–810 g/acre	37%–73%	Low-input perennial grass. Ethanol production depends on development of cellulosic technology. Breeding efforts underway to increase yields. Higher biomass production possible with mixed species of perennial grasses.
Poplar	3700–6000 L/ha, 400–640 g/acre	51%–100%	Fast-growing tree. Ethanol production depends on development of cellulosic technology. Completion of genomic sequencing project will aid breeding efforts to increase yields.
Sweet sorghum	2500–7000 L/ha, 270–750 g/acre	No data	Low-input annual grass. Ethanol production possible using existing technology. Grows in tropical and temperate climates, but highest ethanol yield estimates assume multiple crops per year (possible only in tropical climates). Does not store well.
Corn	3100–4000 L/ha, 330–424 g/acre	10%–20%	High-input annual grass. Used as feedstock for most bioethanol produced in USA. Only kernels can be processed using available technology; development of commercial cellulosic technology would allow stover to be used and increase ethanol yield by 1,100 – 2,000 litres/ha.
Source (except those indicated): *Nature* 444 (7 December 2006): 673–676. [a] – Savings of GHG emissions assuming no land use change (using existing crop lands).			

Reduced Petroleum Imports and Costs

One rationale given for extensive ethanol production in the U.S. is its benefit to energy security, by shifting the need for some foreign-produced oil to domestically produced energy sources. Production of ethanol requires significant energy, but current U.S. production derives most of that energy from coal, natural gas and other sources, rather than oil. Because 66% of oil consumed in the U.S. is imported, compared to a net surplus of coal and just 16% of natural gas (figures from 2006), the displacement of oil-based fuels to ethanol produces a net shift from foreign to domestic U.S. energy sources.

According to a 2008 analysis by Iowa State University, the growth in US ethanol production has caused retail gasoline prices to be US $0.29 to US $0.40 per gallon lower than would otherwise have been the case.

Motorsport

Leon Duray qualified third for the 1927 Indianapolis 500 auto race with an ethanol-fueled car. The IndyCar Series adopted a 10% ethanol blend for the 2006 season, and a 98% blend in 2007.

In drag racing, there are Top Alcohol classes for dragsters and funny cars since the 1970s.

The American Le Mans Series sports car championship introduced E10 in the 2007 season to replace pure gasoline. In the 2008 season, E85 was allowed in the GT class and teams began switching to it.

In 2011, the three national NASCAR stock car series mandated a switch from gasoline to E15, a blend of Sunoco GTX unleaded racing fuel and 15% ethanol.

Australia's V8 Supercar championship uses United E85 for its racing fuel.

Stock Car Brasil Championship runs on neat ethanol, E100.

Ethanol fuel may also be utilized as a rocket fuel. As of 2010, small quantities of ethanol are used in lightweight rocket-racing aircraft.

Replacement Cooking Fuel

A non-profit named Project Gaia seeks to spread the use of ethanol stoves to replace wood, charcoal and kerosene.

Research

Ethanol research focuses on alternative sources, novel catalysts and production processes.

Ethanol plant in Turner County, South Dakota

In 2013, INEOS began initial operation of a bio-ethanol plant from vegetative material and wood waste.

The bacterium E.coli when genetically engineered with cow rumen genes and enzymes can produce ethanol from corn stover.

An alternative technology allows for the production of biodiesel from grain that has already been used to produce ethanol.

Another approach uses feed stocks such as municipal waste, recycled products, rice hulls, sugarcane bagasse, wood chips or switchgrass.

Bibliography

- *J. Goettemoeller; A. Goettemoeller (2007). Sustainable Ethanol: Biofuels, Biorefineries, Cellulosic Biomass, Flex-Fuel Vehicles, and Sustainable Farming for Energy Independence (Brief and comprehensive account of the history, evolution and future of ethanol). Prairie Oak Publishing, Maryville, Missouri. ISBN 978-0-9786293-0-4.*

- *Onuki, Shinnosuke; Koziel, Jacek A.; van Leeuwen, Johannes; Jenks, William S.; Grewell, David; Cai, Lingshuang (June 2008).* Ethanol production, purification, and analysis techniques: a review. *2008 ASABE Annual International Meeting. Providence, Rhode Island. Retrieved February 16, 2013.*

- *The Worldwatch Institute (2007). Biofuels for Transport: Global Potential and Implications for Energy and Agriculture (Global view, includes country study cases of Brazil, China, India and Tanzania). London, UK: Earthscan Publications. ISBN 978-1-84407-422-8.*

Second-generation Biofuels

Second generation biofuels, also known as advanced biofuels, are fuels that can be manufactured from various types of biomass. Biomass is a wide-ranging term meaning any source of organic carbon that is renewed rapidly as part of the carbon cycle. Biomass is derived from plant materials but can also include animal materials.

First generation biofuels are made from the sugars and vegetable oils found in arable crops, which can be easily extracted using conventional technology. In comparison, second generation biofuels are made from lignocellulosic biomass or woody crops, agricultural residues or waste, which makes it harder to extract the required fuel.

Introduction

Second generation biofuel technologies have been developed because first generation biofuels manufacture has important limitations. First generation biofuel processes are useful but limited in most cases: there is a threshold above which they cannot produce enough biofuel without threatening food supplies and biodiversity. Many first generation biofuels depend on subsidies and are not cost competitive with existing fossil fuels such as oil, and some of them produce only limited greenhouse gas emissions savings. When taking emissions from production and transport into account, life-cycle assessment from first generation biofuels frequently approach those of traditional fossil fuels.

Second generation biofuels can help solve these problems and can supply a larger proportion of global fuel supply sustainably, affordably, and with greater environmental benefits.

First generation bioethanol is produced by fermenting plant-derived sugars to ethanol, using a similar process to that used in beer and wine-making. This requires the use of 'food' crops, such as sugar cane, corn, wheat, and sugar beet. These crops are required for food, so, if too much biofuel is made from them, food prices could rise and shortages might be experienced in some countries. Corn, wheat, and sugar beet can also require high agricultural inputs in the form of fertilizers, which limit the greenhouse gas reductions that can be achieved. Biodiesel produced by transesterification from rapeseed oil, palm oil, or other plant oils is also considered a first generation biofuel.

The goal of second generation biofuel processes is to extend the amount of biofuel that can be produced sustainably by using biomass consisting of the residual non-food parts of current crops, such as stems, leaves and husks that are left behind once the food crop has been extracted, as well as other crops that are not used for food purposes (non-food crops), such as switchgrass, grass, jatropha, whole crop maize, miscanthus and cereals that bear little grain, and also industry waste such as woodchips, skins and pulp from fruit pressing, etc.

The problem that second generation biofuel processes are addressing is to extract useful feedstocks from this woody or fibrous biomass, where the useful sugars are locked in by lignin, hemicellulose and cellulose. All plants contain lignin, hemicellulose and cellulose. These are complex carbohydrates (molecules based on sugar). Lignocellulosic ethanol is made by freeing the sugar molecules from cellulose using enzymes, steam heating, or other pre-treatments. These sugars can then be fermented to produce ethanol in the same way as first generation bioethanol production. The by-product of this process is lignin. Lignin can be burned as a carbon neutral fuel to produce heat and power for the processing plant and possibly for surrounding homes and businesses. Thermochemical processes (liquefaction) in hydrothermal media can produce liquid oily products from a wide range of feedstock that has a potential to replace or augment fuels. However, these liquid products fall short of diesel or biodiesel standards. Upgrading liquefaction products through one or many physical or chemical processes may improve properties for use as fuel.

Second Generation Technology

The following subsections describe the main second generation routes currently under development.

Thermochemical Routes

Carbon-based materials can be heated at high temperatures in the absence (pyrolysis) or presence of oxygen, air and/or steam (gasification).

These thermochemical processes both yield a combustible gas and a solid char. The gas can be fermented or chemically synthesised into a range of fuels, including ethanol, synthetic diesel or jet fuel.

There are also lower temperature processes in the region of 150-374 °C, that produce sugars by decomposing the biomass in water with or without additives.

Gasification

Gasification technologies are well established for conventional feedstocks such as coal and crude oil. Second generation gasification technologies include gasification of forest and agricultural residues, waste wood, energy crops and black liquor. Output is normally syngas for further synthesis to e.g. Fischer-Tropsch products including diesel fuel, biomethanol, BioDME (dimethyl ether), gasoline via catalytic conversion of dimethyl ether, or biomethane (synthetic natural gas). Syngas can also be used in heat production and for generation of mechanical and electrical power via gas motors or gas turbines.

Pyrolysis

Pyrolysis is a well established technique for decomposition of organic material at elevated temperatures in the absence of oxygen. In second generation biofuels applications forest and agricultural residues, wood waste and energy crops can be used as feedstock to produce e.g. bio-oil for fuel oil applications. Bio-oil typically requires significant additional treatment to render it suitable as a refinery feedstock to replace crude oil.

Torrefaction

Torrefaction is a form of pyrolysis at temperatures typically ranging between 200-320 °C. Feedstocks and output are the same as for pyrolysis.

Biochemical Routes

Chemical and biological processes that are currently used in other applications are being adapted for second generation biofuels. Biochemical processes typically employ pre-treatment to accelerate the hydrolysis process, which separates out the lignin, hemicellulose and cellulose. Once these ingredients are separated, the cellulose fractions can be fermented into alcohols.

Feedstocks are energy crops, agricultural and forest residues, food industry and municipal biowaste and other biomass containing sugars. Products include alcohols (such as ethanol and butanol) and other hydrocarbons for transportation use.

Types of Biofuel

The following second generation biofuels are under development, although most or all of these biofuels are synthesized from intermediary products such as syngas using methods that are identical in processes involving conventional feedstocks, first generation and second generation biofuels. The distinguishing feature is the technology involved in producing the intermediary product, rather than the ultimate off-take.

A process producing liquid fuels from gas (normally syngas) is called a Gas-to-Liquid (GtL) process. When biomass is the source of the gas production the process is also referred to as Biomass-To-Liquids (BTL).

From Syngas Using Catalysis

- Biomethanol can be used in methanol motors or blended with petrol up to 10-20% without any infrastructure changes.

- BioDME can be produced from Biomethanol using catalytic dehydration or it can be produced directly from syngas using direct DME synthesis. DME can be used in the compression ignition engine.

- Bio-derived gasoline can be produced from DME via high-pressure catalytic condensation reaction. Bio-derived gasoline is chemically indistinguishable from petroleum-derived gasoline and thus can be blended into the U.S. gasoline pool.

- Biohydrogen can be used in fuel cells to produce electricity.

- Mixed Alcohols (i.e., mixture of mostly ethanol, propanol, and butanol, with some pentanol, hexanol, heptanol, and octanol). Mixed alcohols are produced from syngas with several classes of catalysts. Some have employed catalysts similar to those used for methanol. Molybdenum sulfide catalysts were discovered at Dow Chemical and have received considerable attention. Addition of cobalt sulfide to the catalyst formulation was shown to enhance performance. Molybdenum sulfide catalysts have been well studied but have yet to find widespread use. These catalysts have been a focus of efforts at the U.S. Department of Energy's Biomass Program in the Thermochemical Platform. Noble metal catalysts have also been shown to produce mixed alcohols. Most R&D in this area is concentrated in producing mostly ethanol. However, some fuels are marketed as mixed alcohols Mixed alcohols are superior to pure methanol or ethanol, in that the higher alcohols have higher energy content. Also, when blending, the higher alcohols increase compatibility of gasoline and ethanol, which increases water tolerance and decreases evaporative emissions. In addition, higher alcohols have also lower heat of vaporization than ethanol, which is important for cold starts.

- Biomethane (or Bio-SNG) via the Sabatier reaction

From Syngas Using Fischer-Tropsch

The Fischer-Tropsch (FT) process is a Gas-to-Liquid (GtL) process. When biomass is the source of the gas production the process is also referred to as Biomass-To-Liquids (BTL). A disadvantage of this process is the high energy investment for the FT synthesis and consequently, the process is not yet economic.

- FT diesel can be mixed with fossil diesel at any percentage without need for infrastructure change and moreover, synthetic kerosene can be produced

Biocatalysis

- Biohydrogen might be accomplished with some organisms that produce hydrogen directly under certain conditions. Biohydrogen can be used in fuel cells to produce electricity.

- Butanol and Isobutanol via recombinant pathways expressed in hosts such as E. coli and yeast, butanol and isobutanol may be significant products of fermentation using glucose as a carbon and energy source.

- DMF (2,5-Dimethylfuran). Recent advances in producing DMF from fructose and glucose using catalytic biomass-to-liquid process have increased its attractiveness.

Other Processes

- HTU (Hydro Thermal Upgrading) diesel is produced from wet biomass. It can be mixed with fossil diesel in any percentage without need for infrastructure.

- Wood diesel. A new biofuel was developed by the University of Georgia from woodchips. The oil is extracted and then added to unmodified diesel engines. Either new plants are used or planted to replace the old plants. The charcoal byproduct is put back into the soil as a fertilizer. According to the director Tom Adams since carbon is put back into the soil, this biofuel can actually be carbon negative not just carbon neutral. Carbon negative decreases carbon dioxide in the air reversing the greenhouse effect not just reducing it.

Feedstocks

Second generation biofuel feedstocks include cereal and sugar crops, specifically grown energy crops, agricultural and municipal wastes, cultivated and waste oils, and algae. Land use, existing biomass industries and relevant conversion technologies must be considered when evaluating suitability of developing biomass as feedstock for energy.

Energy Crops

Plants are made from lignin, hemicellulose and cellulose; second generation technology uses one, two or all of these components. Common lignocellulosic energy crops include wheat straw, Miscanthus, short rotation coppice poplar and willow. However, each offers different opportunities and no one crop can be considered 'best' or 'worst'.

Municipal Solid Waste

Municipal Solid Waste comprises a very large range of materials, and total waste arisings are increasing. In the UK, recycling initiatives decrease the proportion of waste going straight for disposal, and the level of recycling is increasing each year. However, there remains significant opportunities to convert this waste to fuel via gasification or pyrolysis.

Green Waste

Green waste such as forest residues or garden or park waste may be used to produce biofuel via different routes. Examples include Biogas captured from biodegradable green waste, and gasification or hydrolysis to syngas for further processing to biofuels via catalytic processes.

Black Liquor

Black liquor, the spent cooking liquor from the kraft process that contains concentrated lignin and hemicellulose, may be gasified with very high conversion efficiency and greenhouse gas reduction potential to produce syngas for further synthesis to e.g. biomethanol or BioDME.

Greenhouse Gas Emissions

Lignocellulosic biofuels reduces greenhouse gas emissions with 60-90% when compared with fossil petroleum (Börjesson.P. et al. 2013. Dagens och framtidens hållbara biodrivmedel), which is on par with the better of current biofuels of the first generation, where typical best values currently is 60-80%. In 2010, average savings of biofuels used within EU was 60% (Hamelinck.C. et al. 2013 Renewable energy progress and biofuels sustainability, Report for the European Commission). In 2013, 70% of the biofuels used in Sweden reduced emissions with 66% or higher. (Energimyndigheten 2014. Hållbara biodrivmedel och flytande biobränslen 2013).

Commercial Development

An operating lignocellulosic ethanol production plant is located in Canada, run by Iogen Corporation. The demonstration-scale plant produces around 700,000 litres of bioethanol each year. A commercial plant is under construction. Many further lignocellulosic ethanol plants have been proposed in North America and around the world.

The Swedish specialty cellulose mill Domsjö Fabriker in Örnsköldsvik, Sweden develops a biorefinery using Chemrec's black liquor gasification technology. When commissioned in 2015 the biorefinery will produce 140,000 tons of biomethanol or 100,000 tons of BioDME per year, replacing 2% of Sweden's imports of diesel fuel for transportation purposes. In May 2012 it was revealed that Domsjö pulled out of the project, effectively killing the effort.

In the UK, companies like INEOS Bio and British Airways are developing advanced biofuel refineries, which are due to be built by 2013 and 2014 respectively. Under favourable economic conditions and strong improvements in policy support, NNFCC projections suggest advanced biofuels could meet up to 4.3 per cent of the UK's transport fuel by 2020 and save 3.2 million tonnes of CO_2 each year, equivalent to taking nearly a million cars off the road.

Helsinki, Finland, 1 February 2012 - UPM is to invest in a biorefinery producing biofuels from crude tall oil in Lappeenranta, Finland. The industrial scale investment is the first of its kind globally. The biorefinery will produce annually approximately 100,000 tonnes of advanced second generation biodiesel for transport. Construction of the biorefinery will begin in the summer of 2012 at UPM's Kaukas mill site and be completed in 2014. UPM's total investment will amount to approximately EUR 150 million.

Calgary, Alberta, 30 April 2012 – Iogen Energy Corporation has agreed to a new plan with its joint owners Royal Dutch Shell and Iogen Corporation to refocus its strategy and activities. Shell continues to explore multiple pathways to find a commercial solution for the production of advanced biofuels on an industrial scale, but the company will NOT pursue the project it has had under development to build a larger scale cellulosic ethanol facility in southern Manitoba.

"Drop-in" Biofuels

So-called "drop-in" biofuels can be defined as "liquid bio-hydrocarbons that are functionally equivalent to petroleum fuels and are fully compatible with existing petroleum infrastructure".

There is considerable interest in developing advanced biofuels that can be readily integrated in the existing petroleum fuel infrastructure - i.e. "dropped-in" - particularly by sectors such as aviation, where there are no real alternatives to sustainably produced biofuels for low carbon emitting fuel sources. Drop-in biofuels by definition should be fully fungible and compatible with the large existing "petroleum-based" infrastructure.

According to a recent report published by the IEA Bioenergy Task 39, entitled "The potential and challenges of drop-in biofuels", there are several ways to produce drop-in biofuels that are functionally equivalent to petroleum-derived transportation fuel blendstocks. These are discussed within three major sections of the full report and include:

- oleochemical processes, such as the hydroprocessing of lipid feedstocks obtained from oilseed crops, algae or tallow;

- thermochemical processes, such as the thermochemical conversion of biomass to fluid intermediates (gas or oil) followed by catalytic upgrading and hydroprocessing to hydrocarbon fuels; and

- biochemical processes, such as the biological conversion of biomass (sugars, starches or lignocellulose-derived feedstocks) to longer chain alcohols and hydrocarbons.

A fourth category is also briefly described that includes "hybrid" thermochemical/biochemical technologies such as fermentation of synthesis gas and catalytic reforming of sugars/carbohydrates.

The report concludes by stating:

"Tremendous entrepreneurial activity to develop and commercialize drop-in biofuels from aquatic and terrestrial feedstocks has taken place over the past several years. However, despite these efforts, drop-in biofuels represent only a small percentage (around 2%) of global biofuel markets. (...) Due to the increased processing and resource requirements (e.g., hydrogen and catalysts) needed to make drop-in biofuels as compared to conventional biofuels, large scale production of cost-competitive drop-in biofuels is not expected to occur in the near to midterm. Rather, dedicated policies to promote development and commercialization of these fuels will be needed before they become significant contributors to global biofuels production. Currently, no policies (e.g., tax breaks, subsidies etc.) differentiate new, more fungible and infrastructure ready drop-in type biofuels from less infrastructure compatible oxygenated biofuels. (...) Thus, while tremendous technical progress has been made in developing and improving the various routes to drop-in fuels, supportive policies directed specifically towards the further development of drop-in biofuels are likely to be needed to ensure their future commercial success"

Sustainable Biofuel

Biofuels, in the form of liquid fuels derived from plant materials, are entering the market, driven by factors such as oil price spikes and the need for increased energy security. However, many of the biofuels that are currently being supplied have been criticised for their adverse impacts on the natural environment, food security, and land use.

The challenge is to support biofuel development, including the development of new cellulosic technologies, with responsible policies and economic instruments to help ensure that biofuel commercialization is sustainable. Responsible commercialization of biofuels represents an opportunity to enhance sustainable economic prospects in Africa, Latin America and Asia.

Biofuels have a limited ability to replace fossil fuels and should not be regarded as a 'silver bullet' to deal with transport emissions. However, they offer the prospect of increased market competition and oil price moderation. A healthy supply of alternative energy sources will help to combat gasoline price spikes and reduce dependency on fossil fuels, especially in the transport sector. Using transportation fuels more efficiently is also an integral part of a sustainable transport strategy.

Biofuel Options

Biofuel development and use is a complex issue because there are many biofuel options which are available. Biofuels, such as ethanol and biodiesel, are currently produced from the products of conventional food crops such as the starch, sugar and oil feedstocks from crops that include wheat, maize, sugar cane, palm oil and oilseed rape. Some researchers fear that a major switch to biofuels from such crops would create a direct competition with their use for food and animal feed, and claim that in some parts of the world the economic consequences are already visible, other researchers look at the land available and the enormous areas of idle and abandoned land and claim that there is room for a large proportion of biofuel also from conventional crops.

Second generation biofuels are now being produced from a much broader range of feedstocks including the cellulose in dedicated energy crops (perennial grasses such as switchgrass and Miscanthus giganteus), forestry materials, the co-products from food production, and domestic vegetable waste. Advances in the conversion processes will improve the sustainability of biofuels, through better efficiencies and reduced environmental impact of producing biofuels, from both existing food crops and from cellulosic sources.

In 2007, Ronald Oxburgh suggested in The Courier-Mail that production of biofuels could be either responsible or irresponsible and had several trade-offs: "Produced responsibly they are a sustainable energy source that need not divert any land from growing food nor damage the environment; they can also help solve the problems of the waste generated by Western society; and they can create jobs for the poor where previously were none. Produced irresponsibly, they at best offer no climate benefit and, at worst, have detrimental social and environmental consequences. In other words, biofuels are pretty much like any other product. In 2008 the Nobel prize-winning chemist Paul J. Crutzen published findings that the release of nitrous oxide (N_2O) emissions in the production of biofuels means that they contribute more to global warming than the fossil fuels they replace.

According to the Rocky Mountain Institute, sound biofuel production practices would not hamper food and fibre production, nor cause water or environmental problems, and would enhance soil fertility. The selection of land on which to grow the feedstocks is a critical component of the ability of biofuels to deliver sustainable solutions. A key consideration is the minimisation of biofuel competition for prime cropland.

Plants Used as Sustainable Biofuel

Sugarcane in Brazil

Brazil's production of ethanol fuel from sugarcane dates back to the 1970s, as a governmental response to the 1973 oil crisis. Brazil is considered the biofuel industry leader and the world's first sustainable biofuels economy. In 2010 the U.S. Environmental Protection Agency designated Brazilian sugarcane ethanol as an advanced biofuel due to EPA's estimated 61% reduction of total life cycle greenhouse gas emissions, including direct indirect land use change emissions. Brazil sugarcane ethanol fuel program success and sustainability is based on the most efficient agricultural technology for sugarcane cultivation in the world, uses modern equipment and cheap sugar cane as feedstock, the residual cane-waste (bagasse) is used to process heat and power, which results in a very competitive price and also in a high energy balance (output energy/input energy), which varies from 8.3 for average conditions to 10.2 for best practice production.

Mechanized harvesting of sugarcane, Piracicaba, São Paulo, Brazil.

A report commissioned by the United Nations, based on a detailed review of published research up to mid-2009 as well as the input of independent experts world-wide, found that ethanol from sugar cane as produced in Brazil *"in some circumstances does better than just "zero emission". If grown and processed correctly, it has negative emission, pulling CO_2 out of the atmosphere, rather than adding it.* In contrast, the report found that U.S. use of maize for biofuel is less efficient, as sugarcane can lead to emissions reductions of between 70% and well over 100% when substituted for gasoline. Several other studies have shown that sugarcane-based ethanol reduces greenhouse gases by 86 to 90% if there is no significant land use change.

Cosan's Costa Pinto sugar cane mill and ethanol distillery plant at Piracicaba, São Paulo, Brazil.

In another study commissioned by the Dutch government in 2006 to evaluate the sustainability of Brazilian bioethanol concluded that there is sufficient water to supply all foreseeable long-term water requirements for sugarcane and ethanol production. This evaluation also found that consumption of agrochemicals for sugar cane production is lower than in citric, corn, coffee and soybean cropping. The study found that development of resistant sugar cane varieties is a crucial aspect of disease and pest control and is one of the primary objectives of Brazil's cane genetic improvement programs. Disease control is one of the main reasons for the replacement of a commercial variety of sugar cane.

Another concern is the fact that sugarcane fields are traditionally burned just before harvest to avoid harm to the workers, by removing the sharp leaves and killing snakes and other harmful animals, and also to fertilize the fields with ash. Mechanization will reduce pollution from burning fields and has higher productivity than people, and due to mechanization the number of temporary workers in the sugarcane plantations has already declined. By the 2008 harvest season, around 47% of the cane was collected with harvesting machines.

Regarding the negative impacts of the potential direct and indirect effect of land use changes on carbon emissions, the study commissioned by the Dutch government concluded that "it is very difficult to determine the indirect effects of further land use for sugar cane production (i.e. sugar cane replacing another crop like soy or citrus crops, which in turn causes additional soy plantations replacing pastures, which in turn may cause deforestation), and also not logical to attribute all these soil carbon losses to sugar cane". The Brazilian agency Embrapa estimates that there is enough agricultural land available to increase at least 30 times the existing sugarcane plantation without endangering sensible ecosystems or taking land destined for food crops. Most future growth is expected to take place on abandoned pasture lands, as it has been the historical trend in São Paulo state. Also, productivity is expected to improve even further based on current biotechnology research, genetic improvement, and better agronomic practices, thus contributing to reduce land demand for future sugarcane cultures.

Location of environmentally valuable areas with respect to sugarcane plantations. São Paulo, located in the Southeast Region of Brazil, concentrates two-thirds of sugarcane cultures.

Another concern is the risk of clearing rain forests and other environmentally valuable land for sugarcane production, such as the Amazonia, the Pantanal or the Cerrado. Embrapa has rebutted this concern explaining that 99.7% of sugarcane plantations are located at least 2,000 km from the Amazonia, and expansion during the last 25 years took place in the Center-South region, also far away from the Amazonia, the Pantanal or the Atlantic forest. In São Paulo state growth took place in abandoned pasture lands.

The impact assessment commissioned by the Dutch government supported this argument.

In order to guarantee a sustainable development of ethanol production, in September 2009 the government issued by decree a countrywide agroecological land use zoning to restrict sugarcane growth in or near environmentally sensitive areas. According to the new criteria, 92.5% of the Brazilian territory is not suitable for sugarcane plantation. The government considers that the suitable areas are more than enough to meet the future demand for ethanol and sugar in the domestic and international markets foreseen for the next decades.

Regarding the food vs fuel issue, a World Bank research report published on July 2008 found that *"Brazil's sugar-based ethanol did not push food prices appreciably higher"*. This research paper also concluded that Brazil's sugar cane–based ethanol has not raised sugar prices significantly. An economic assessment report also published in July 2008 by the OECD agrees with the World Bank report regarding the negative effects of subsidies and trade restrictions, but found that the impact of biofuels on food prices are much smaller. A study by the Brazilian research unit of the Fundação Getúlio Vargas regarding the effects of biofuels on grain prices concluded that the major driver behind the 2007-2008 rise in food prices was speculative activity on futures markets under conditions of increased demand in a market with low grain stocks. The study also concluded that there is no correlation between Brazilian sugarcane cultivated area and average grain prices, as on the contrary, the spread of sugarcane was accompanied by rapid growth of grain crops in the country.

Jatropha

India and Africa

Jatropha gossipifolia in Hyderabad, India.

Crops like Jatropha, used for biodiesel, can thrive on marginal agricultural land where many trees and crops won't grow, or would produce only slow growth yields. Jatropha cultivation provides benefits for local communities:

Cultivation and fruit picking by hand is labour-intensive and needs around one person per hectare. In parts of rural India and Africa this provides much-needed jobs - about 200,000 people worldwide now find employment through jatropha. Moreover, villagers often find that they can grow other crops in the shade of the trees. Their communities will avoid importing expensive diesel and there will be some for export too.

Cambodia

Cambodia has no proven fossil fuel reserves, and is almost completely dependent on imported diesel fuel for electricity production. Consequently, Cambodians face an insecure supply and pay some of the highest energy prices in the world. The impacts of this are widespread and may hinder economic development.

Biofuels may provide a substitute for diesel fuel that can be manufactured locally for a lower price, independent of the international oil price. The local production and use of biofuel also offers other benefits such as improved energy security, rural development opportunities and environmental benefits. The Jatropha curcas species appears to be a particularly suitable source of biofuel as it already grows commonly in Cambodia. Local sustainable production of biofuel in Cambodia, based on the Jatropha or other sources, offers good potential benefits for the investors, the economy, rural communities and the environment.

Mexico

Jatropha is native to Mexico and Central America and was likely transported to India and Africa in the 1500s by Portuguese sailors convinced it had medicinal uses. In 2008, recognizing the need to diversify its sources of energy and reduce emissions, Mexico passed a law to push developing biofuels that don't threaten food security and the agriculture ministry has since identified some 2.6 million hectares (6.4 million acres) of land with a high potential to produce jatropha. The Yucatán Peninsula, for instance, in addition to being a corn producing region, also contains abandoned sisal plantations, where the growing of Jatropha for biodiesel production would not displace food.

On April 1, 2011 Interjet completed the first Mexican aviation biofuels test flight on an Airbus A320. The fuel was a 70:30 traditional jet fuel biojet blend produced from Jatropha oil provided by three Mexican producers, Global Energías Renovables (a wholly owned subsidiary of U.S.-based Global Clean Energy Holdings, Bencafser S.A. and Energy JH S.A. Honeywell's UOP processed the oil into Bio-SPK (Synthetic Paraffinic Kerosene) . Global Energías Renovables operates the largest Jatropha farm in the Americas.

On August 1, 2011 Aeromexico, Boeing, and the Mexican Government participated in the first biojet powered transcontinental flight in aviation history. The flight from Mexico City to Madrid used a blend of 70 percent traditional fuel and 30 percent biofuel (aviation biofuel). The biojet was produced entirely from Jatropha oil.

Pongamia Pinnata in Australia and India

Pongamia pinnata is a legume native to Australia, India, Florida (USA) and most tropical regions, and is now being invested in as an alternative to Jatropha for areas such as Northern Australia, where Jatropha is classed as a noxious weed. Commonly known as simply 'Pongamia', this tree is currently being commercialised in Australia by Pacific Renewable Energy, for use as a Diesel replacement for running in modified Diesel engines or for conversion to Biodiesel using 1st or 2nd Generation Biodiesel techniques, for running in unmodified Diesel engines.

Pongamia pinnata seeds in Brisbane, Australia.

Sweet Sorghum in India

Sweet sorghum overcomes many of the shortcomings of other biofuel crops. With sweet sorghum, only the stalks are used for biofuel production, while the grain is saved for food or livestock feed. It is not in high demand in the global food market, and thus has little impact on food prices and food security. Sweet sorghum is grown on already-farmed drylands that are low in carbon storage capacity, so concerns about the clearing of rainforest do not apply. Sweet sorghum is easier and cheaper to grow than other biofuel crops in India and does not require irrigation, an important consideration in dry areas. Some of the Indian sweet sorghum varieties are now grown in Uganda for ethanol production.

A study by researchers at the International Crops Research Institute for the Semi-Arid Tropics (ICRISAT) found that growing sweet sorghum instead of grain sorghum could increase farmers incomes by US$40 per hectare per crop because it can provide food, feed and fuel. With grain sorghum currently grown on over 11 million hectares (ha) in Asia and on 23.4 million ha in Africa, a switch to sweet sorghum could have a considerable economic impact.

International Collaboration on Sustainable Biofuels

Roundtable on Sustainable Biofuels

Public attitudes and the actions of key stakeholders can play a crucial role in realising the potential of sustainable biofuels. Informed discussion and dialogue, based both on scientific research and an understanding of public and stakeholder views, is important.

The Roundtable on Sustainable Biofuels is an international initiative which brings together farmers, companies, governments, non-governmental organizations, and scientists who are interested in the sustainability of biofuels production and distribution. During 2008, the Roundtable used meetings, teleconferences, and online discussions to develop a series of principles and criteria for sustainable biofuels production.

In 2008, the Roundtable for Sustainable Biofuels released its proposed standards for sustainable biofuels. This includes 12 principles:

1. "Biofuel production shall follow international treaties and national laws regarding such things as air quality, water resources, agricultural practices, labor conditions, and more.

2. Biofuels projects shall be designed and operated in participatory processes that involve all relevant stakeholders in planning and monitoring.

3. Biofuels shall significantly reduce greenhouse gas emissions as compared to fossil fuels. The principle seeks to establish a standard methodology for comparing greenhouse gases (GHG) benefits.

4. Biofuel production shall not violate human rights or labor rights, and shall ensure decent work and the well-being of workers.

5. Biofuel production shall contribute to the social and economic development of local, rural and indigenous peoples and communities.

6. Biofuel production shall not impair food security.

7. Biofuel production shall avoid negative impacts on biodiversity, ecosystems and areas of high conservation value.

8. Biofuel production shall promote practices that improve soil health and minimize degradation.

9. Surface and groundwater use will be optimized and contamination or depletion of water resources minimized.

10. Air pollution shall be minimized along the supply chain.

11. Biofuels shall be produced in the most cost-effective way, with a commitment to

improve production efficiency and social and environmental performance in all stages of the biofuel value chain.

12. Biofuel production shall not violate land rights".

In April 2011, the Roundtable on Sustainable Biofuels launched a set of comprehensive sustainability criteria - the "RSB Certification System." Biofuels producers that meet to these criteria are able to show buyers and regulators that their product has been obtained without harming the environment or violating human rights.

Sustainable Biofuels Consensus

The Sustainable Biofuels Consensus is an international initiative which calls upon governments, the private sector, and other stakeholders to take decisive action to ensure the sustainable trade, production, and use of biofuels. In this way biofuels may play a key role in energy sector transformation, climate stabilization, and resulting worldwide revitalisation of rural areas.

The Sustainable Biofuels Consensus envisions a "landscape that provides food, fodder, fiber, and energy, which offers opportunities for rural development; that diversifies energy supply, restores ecosystems, protects biodiversity, and sequesters carbon".

Better Sugarcane Initiative / Bonsucro

In 2008, a multi-stakeholder process was initiated by the World Wildlife Fund and the International Finance Corporation, the private development arm of the World Bank, bringing together industry, supply chain intermediaries, end-users, farmers and civil society organisations to develop standards for certifying the derivative products of sugar cane, one of which is ethanol fuel.

The Bonsucro standard is based around a definition of sustainability which is founded on five principles:

1. Obey the law

2. Respect human rights and labour standards

3. Manage input, production and processing efficiencies to enhance sustainability

4. Actively manage biodiversity and ecosystem services

5. Continuously improve key areas of the business

Biofuel producers that wish to sell products marked with the Bonsucro standard must both ensure that they product to the Production Standard, and that their downstream buyers meet the Chain of Custody Standard. In addition, if they wish to sell to the European market and count against the EU Renewable Energy Directive, then they must adhere to the Bonsucro EU standard, which includes specific greenhouse gas calculations following European Commission calculation guidelines.

Sustainability Standards

Several countries and regions have introduced policies or adopted standards to promote sustainable biofuels production and use, most prominently the European Union and the United States. The 2009 EU Renewable Energy Directive, which requires 10 percent of transportation energy from renewable energy by 2020, is the most comprehensive mandatory sustainability standard in place as of 2010. The Directive requires that the lifecycle greenhouse gas emissions of biofuels consumed be at least 50 percent less than the equivalent emissions from gasoline or diesel by 2017 (and 35 percent less starting in 2011). Also, the feedstocks for biofuels "should not be harvested from lands with high biodiversity value, from carbon-rich or forested land, or from wetlands".

As with the EU, the U.S. Renewable Fuel Standard (RFS) and the California Low Carbon Fuel Standard (LCFS) both require specific levels of lifecycle greenhouse gas reductions compared to equivalent fossil fuel consumption. The RFS requires that at least half of the biofuels production mandated by 2022 should reduce lifecycle emissions by 50 percent. The LCFS is a performance standard that calls for a minimum of 10 percent emissions reduction per unit of transport energy by 2020. Both the U.S. and California standards currently address only greenhouse gas emissions, but California plans to "expand its policy to address other sustainability issues associated with liquid biofuels in the future".

In 2009, Brazil also adopted new sustainability policies for sugarcane ethanol, including "zoning regulation of sugarcane expansion and social protocols".

Oil Price Moderation

Biofuels offer the prospect of real market competition and oil price moderation. According to the Wall Street Journal, crude oil would be trading 15 per cent higher and gasoline would be as much as 25 per cent more expensive, if it were not for biofuels. A healthy supply of alternative energy sources will help to combat gasoline price spikes.

Sustainable Transport

Biofuels have a limited ability to replace fossil fuels and should not be regarded as a 'silver bullet' to deal with transport emissions. Biofuels on their own cannot deliver a sustainable transport system and so must be developed as part of an integrated approach, which promotes other renewable energy options and energy efficiency, as well as reducing the overall energy demand and need for transport. Consideration needs to be given to the development of hybrid and fuel cell vehicles, public transport, and better town and rural planning.

In December 2008 an Air New Zealand jet completed the world's first commercial aviation test flight partially using jatropha-based fuel. More than a dozen performance tests were undertaken in the two-hour test flight which departed from Auckland International Airport. A biofuel blend of 50:50 jatropha and Jet A1 fuel was used to power one of the Boeing 747-400's Rolls-Royce RB211 engines. Air New Zealand set several criteria for its jatropha, requiring that "the land it came from was neither forest nor virgin grassland in the previous 20 years, that the soil and climate it came from is not suitable for the majority of food crops and that the farms are rain fed and not mechanically irrigated". The company has also set general sustainability criteria, saying that such biofuels must not compete with food resources, that they must be as good as traditional jet fuels, and that they should be cost competitive.

In January 2009, Continental Airlines used a sustainable biofuel to power a commercial aircraft for the first time in North America. This demonstration flight marks the first sustainable biofuel demonstration flight by a commercial carrier using a twin-engined aircraft, a Boeing 737-800, powered by CFM International CFM56-7B engines. The biofuel blend included components derived from algae and jatropha plants. The algae oil was provided by Sapphire Energy, and the jatropha oil by Terasol Energy.

In March 2011, Yale University research showed significant potential for sustainable aviation fuel based on jatropha-curcas. According to the research, if cultivated properly, "jatropha can deliver many benefits in Latin America and greenhouse gas reductions of up to 60 percent when compared to petroleum-based jet fuel". Actual farming conditions in Latin America were assessed using sustainability criteria developed by the Roundtable on Sustainable Biofuels. Unlike previous research, which used theoretical inputs, the Yale team conducted many interviews with jatropha farmers and used "field measurements to develop the first comprehensive sustainability analysis of actual projects".

As of June 2011, revised international aviation fuel standards officially allow commercial airlines to blend conventional jet fuel with up to 50 percent biofuels. The renewable fuels "can be blended with conventional commercial and military jet fuel through requirements in the newly issued edition of ASTM D7566, Specification for Aviation Turbine Fuel Containing Synthesized Hydrocarbons".

In December 2011, the FAA awarded $7.7 million to eight companies to advance the development of commercial aviation biofuels, with a special focus on alcohol to jet fuel. The FAA is assisting in the development of a sustainable fuel (from alcohols, sugars, biomass, and organic matter such as pyrolysis oils) that can be "dropped in" to aircraft without changing current practices and infrastructure. The research will test how the new fuels affect engine durability and quality control standards.

GreenSky London, a biofuels plant under construction in 2014, will take in some 500,000 tonnes of municipal rubbish and change the organic component into 60,000 tonnes of jet fuel, and 40 megawatts of power. By the end of 2015, all British Airways flights from London City Airport will be fuelled by waste and rubbish discarded by London residents.

Dry Animal Dung Fuel

Dry animal dung fuel (or dry manure fuel) is animal feces that has been dried in order to be used as a fuel source. It is used as a fuel in many countries around the world. Using dry manure as a fuel source is an example of reuse of excreta. A disadvantage of using this kind of fuel is increased air pollution.

Stirling-Motor powered with cow dung in the Technical Collection Hochhut in Frankfurt on Main

Dry Dung and Moist Dung

Dry dung is more commonly used than moist dung, because it burns more easily. Dry manure is typically defined as having a moisture content less than 30 percent.

Benefits

The benefits of using dry animal dung include:

- Cheaper than most modern fuels
- Efficient
- Alleviates local pressure on wood resources
- Readily available - short walking time required to collect fuel

- No cash outlays necessary for purchase (can be exchanged for other products)

- Less environmental pollution

- Safer disposal of animal dung

- Sustainable and renewable energy source

The M.N. Yavari, of Peru built by Thames Iron Works, London in 1861-62 had a Watt steam engine
(powered by dried lama dung) until 1914

Countries

Drying cow dung fuel

Africa

- In Egypt dry animal dung (from cows & buffaloes) is mixed with straw or crop
residues to make dry fuel called "Gella" or "Jilla" dung cakes in modern times
and ""khoroshtof"" in medieval times. Ancient Egyptians used the dry animal
dung as a source of fuel. Dung cakes and building crop residues were the source
of 76.4% of gross energy consumed in Egypt's rural areas during the 1980s.
Temperatures of dung-fueled fires in an experiment on Egyptian village-made
dung cake fuel produced

Egyptian women making "Gella" dry animal dung fuel

""a maximum of 640 degrees C in 12 minutes, falling to 240 degrees C after 25 minutes and 100 degrees C after 46 minutes. These temperatures were obtained without refueling and without bellows etc.""

Also, camel dung is used as fuel in Egypt.

- Lisu is the cakes of dry cow dung fuel in Lesotho

Huts in a village near Maseru, Lesotho. The fuel being used on the fire is dried cattle dung

- Mali

Asia

- Russians dry animal dung is known as ""Kiziak"" which is made by collecting dried animal dung on the steppe, wetting it in water then mixing it with straw then making it in discs which were then dried in the sun. It was used as a source of fuel for the winter and, throughout the summer.

- France in Maison du Marais poitevin in Coulon there is a demonstration of traditional usage of dry dung fuel.

Dung cooking fire. Pushkar India.

- China

Water buffalo dung fuel drying on a wall in a Hani ethnic
minority village in Yuanyang county, Yunnan, China

- Nepal

U.S. soldiers patrolling outside a qalat covered in caked and dried cow dung in an Afghani village

- Iran since prehistoric time to modern eras

- In India dry buffalo dung is used as fuel and it is sometimes a sacred practice to use cow dung fuel in some areas in India. Cow dung is known as ""Gomaya"" or ""Komaya"" in India. Dry animal dung cakes are called Upla in Hindi.

- In Pakistan cow/buffalo dung is used as fuel.

- Bangladesh dry cow dung fuel is called Ghunte.

- Afghanistan

- Kyrgyz Republic Dung is used in specially designed home stoves, which vent to the outside

Cow dung fuel was burnt on the Gauchar's Historical Field, India to gauge the direction of air currents

The Americas

- Early European settlers on the Great Plains of the United States used dried buffalo manure as a fuel. They called it buffalo chips.

- American officials in Texas are studying using dry cow dung as a fuel

- Pueblo Indians used dry animal dung as a fuel

- In Peru, the Yavari steam ship was fueled by Lama dung fuel for several decades.

- Dry dung can be used in the production of celluloid for film.

Human Feces

Human feces can in principal also be dried and used as a fuel source if they are collected in a type of dry toilet, for example an incinerating toilet. Since 2011, the Bill & Melinda Gates Foundation is supporting the development of such toilets as part of their "Reinvent the Toilet Challenge" to promote safer, more effective ways to treat human excreta. The omni-processor is another example of using human feces contained in faecal sludge or sewage sludge as a fuel source.

History

Dry animal dung was used from prehistoric times, including in Ancient Persia and Ancient Egypt. In Equatorial Guinea archaeological evidence has been found of the practice and biblical records indicate animal and human dung were used as fuel.

References

- The Royal Society (January 2008). Sustainable biofuels: prospects and challenges, ISBN 978-0-85403-662-2.

- The Royal Society (January 2008). Sustainable biofuels: prospects and challenges, ISBN 978-0-85403-662-2.

- Inslee, Jay; Bracken Hendricks (2007). "6. Homegrown Energy". Apollo's Fire. Island Press, Washington, D.C. pp. 153–155, 160–161. ISBN 978-1-59726-175-3.

- "Publications - International Resource Panel". Archived from the original on 1 January 2016. Retrieved 30 May 2015.

- EIA. "Ethanol FAQ". How much ethanol is in gasoline, and how does it affect fuel economy?. Retrieved 23 February 2016.

- IEA-ETSAP. "Ethanol internal combustion engines" (PDF). Retrieved 23 February 2016.

- Pentland, William (1 February 2014). "Is Ethanol Eating Your Car's Engine?". Forbes magazine. Retrieved 23 February 2016.

- Voelcker, John (2016-06-14). "Nissan takes a different approach to fuel cells: ethanol". Green Car Reports. Retrieved 2016-06-16.

- "World Energy Outlook 2006" (PDF). Worldenergyoutlook.org. Archived from the original (PDF) on 28 September 2007. Retrieved 20 January 2015.

- "Mechanics see ethanol damaging small engines". msnbc.com. Retrieved 20 January 2015.

- "Ethanol Producer Magazine – The Latest News and Data About Ethanol Production". Ethanolproducer.com. Retrieved 20 January 2015.

- "01.26.2006 - Ethanol can replace gasoline with significant energy savings, comparable impact on greenhouse gases". Berkeley.edu. Retrieved 20 January 2015.

- "Biofuels Deemed a Greenhouse Threat". The New York Times. Retrieved 20 January 2015.

- Joseph Fargione. "Land Clearing and the Biofuel Carbon Debt". Sciencemag.org. Retrieved 20 January 2015.

- Brown, Robert; Jennifer Holmgren. "Fast Pyrolysis and Bio-Oil Upgrading" (PDF). Retrieved 15 March 2012.

- "Alternative & Advanced Fuels". US Department of Energy. Retrieved 7 March 2012.

- Jessica, Ebert. "Breakthroughs in Green Gasoline Production". Biomass Magazine. Retrieved 14 August 2012.

- Biofuels Digest (2011-05-16). "Jatropha blooms again: SG Biofuels secures 250K acres for hybrids". Biofuels Digest. Retrieved 2012-03-08.

- Biofuels Magazine (2011-04-11). "Energy Farming Methods Mature, Improve". Biofuels Magazine. Retrieved 2012-03-08.

- Cocchi, Maurizio (December 2011). "Global Wood Pellet Industry Market and Trade Study" (PDF). IEA Task 40. Retrieved 1 June 2012.

- Frederick, Paul. "2012 VT Wood Chip & Pellet Heating Conference" (PDF). Biomass Energy Resource Center.Retrieved 23 January 2012.

Various Applications of Biomass

The applications of biomass dealt within this chapter are biogas, biochar, bioliquids, blue carbon, straw and woodchips. The mixture of gases produced by the breaking of organic matter is known as biogas. Biochar is used as a product that is added to soil to improve it. The various aspects elucidated in this text are of vital importance, and provides a better understanding of biomass.

Biogas

Pipes carrying biogas (foreground), natural gas and condensate

Biogas typically refers to a mixture of different gases produced by the breakdown of organic matter in the absence of oxygen. Biogas can be produced from raw materials such as agricultural waste, manure, municipal waste, plant material, sewage, green waste or food waste. Biogas is a renewable energy source and in many cases exerts a very small carbon footprint.

Biogas can be produced by anaerobic digestion with anaerobic organisms, which digest material inside a closed system, or fermentation of biodegradable materials.

Biogas is primarily methane (CH_4) and carbon dioxide (CO_2) and may have small amounts of hydrogen sulfide (H_2S), moisture and siloxanes. The gases methane, hydrogen, and carbon monoxide (CO) can be combusted or oxidized with oxygen. This energy release allows biogas to be used as a fuel; it can be used for any heating purpose, such as cooking. It can also be used in a gas engine to convert the energy in the gas into electricity and heat.

Biogas can be compressed, the same way natural gas is compressed to CNG, and used to power motor vehicles. In the UK, for example, biogas is estimated to have the potential to replace around 17% of vehicle fuel. It qualifies for renewable energy subsidies in some parts of the world. Biogas can be cleaned and upgraded to natural gas standards, when it becomes bio-methane. Biogas is considered to be a renewable resource because its production-and-use cycle is continuous, and it generates no net carbon dioxide. Organic material grows, is converted and used and then regrows in a continually repeating cycle. From a carbon perspective, as much carbon dioxide is absorbed from the atmosphere in the growth of the primary bio-resource as is released when the material is ultimately converted to energy.

Production

Biogas production in rural Germany

Biogas is produced as landfill gas (LFG), which is produced by the breakdown of Biodegradable waste inside a landfill due to chemical reactions and microbes, or as digested gas, produced inside an anaerobic digester. A *biogas plant* is the name often given to an anaerobic digester that treats farm wastes or energy crops. It can be produced using anaerobic digesters (air-tight tanks with different configurations). These plants can be fed with energy crops such as maize silage or biodegradable wastes including sewage sludge and food waste. During the process, the microorganisms transform biomass waste into biogas (mainly methane and carbon dioxide) and digestate. The biogas is a renewable energy that can be used for heating, electricity, and many other operations that use a reciprocating internal combustion engine, such as GE Jenbacher or Caterpillar gas engines. Other internal combustion engines such as gas turbines are suitable for the conversion of biogas into both electricity and heat. The digestate is the remaining inorganic matter that was not transformed into biogas. It can be used as an agricultural fertiliser.

There are two key processes: mesophilic and thermophilic digestion which is dependent on temperature. In experimental work at University of Alaska Fairbanks, a 1000-litre digester using psychrophiles harvested from "mud from a frozen lake in Alaska" has produced 200–300 liters of methane per day, about 20%–30% of the output from digesters in warmer climates.

Dangers

The dangers of biogas are mostly similar to those of natural gas, but with an additional risk from the toxicity of its hydrogen sulfide fraction. Biogas can be explosive when mixed in the ratio of one part biogas to 8-20 parts air. Special safety precautions have to be taken for entering an empty biogas digester for maintenance work.

It is important that a biogas system never has negative pressure as this could cause an explosion. Negative gas pressure can occur if too much gas is removed or leaked; Because of this biogas should not be used at pressures below one column inch of water, measured by a pressure gauge.

Frequent smell checks must be performed on a biogas system. If biogas is smelled anywhere windows and doors should be opened immediately. If there is a fire the gas should be shut off at the gate valve of the biogas system.

Landfill Gas

Landfill gas is produced by wet organic waste decomposing under anaerobic conditions in a biogas.

The waste is covered and mechanically compressed by the weight of the material that is deposited above. This material prevents oxygen exposure thus allowing anaerobic microbes to thrive. This gas builds up and is slowly released into the atmosphere if the site has not been engineered to capture the gas. Landfill gas released in an uncontrolled way can be hazardous since it can become explosive when it escapes from the landfill and mixes with oxygen. The lower explosive limit is 5% methane and the upper is 15% methane.

The methane in biogas is 20 times more potent a greenhouse gas than carbon dioxide. Therefore, uncontained landfill gas, which escapes into the atmosphere may significantly contribute to the effects of global warming. In addition, volatile organic compounds (VOCs) in landfill gas contribute to the formation of photochemical smog.

Technical

Biochemical oxygen demand (BOD) is a measure of the amount of oxygen required by aerobic micro-organisms to decompose the organic matter in a sample of water. Knowing the energy density of the material being used in the biodigester as well as the BOD for the liquid discharge allows for the calculation of the daily energy output from a biodigester.

Another term related to biodigesters is effluent dirtiness, which tells how much organic material there is per unit of biogas source. Typical units for this measure are in mg BOD/litre. As an example, effluent dirtiness can range between 800–1200 mg BOD/litre in Panama.

From 1 kg of decommissioned kitchen bio-waste, 0.45 m³ of biogas can be obtained. The price for collecting biological waste from households is approximately €70 per ton.

Composition

Typical composition of biogas		
Compound	Formula	%
Methane	CH_4	50–75
Carbon dioxide	CO_2	25–50
Nitrogen	N_2	0–10
Hydrogen	H_2	0–1
Hydrogen sulfide	H_2S	0–3
Oxygen	O_2	0–0.5
Source: *www.kolumbus.fi, 2007*		

The composition of biogas varies depending upon the origin of the anaerobic digestion process. Landfill gas typically has methane concentrations around 50%. Advanced waste treatment technologies can produce biogas with 55%–75% methane, which for reactors with free liquids can be increased to 80%-90% methane using in-situ gas purification techniques. As produced, biogas contains water vapor. The fractional volume of water vapor is a function of biogas temperature; correction of measured gas volume for water vapor content and thermal expansion is easily done via simple mathematics which yields the standardized volume of dry biogas.

In some cases, biogas contains siloxanes. They are formed from the anaerobic decomposition of materials commonly found in soaps and detergents. During combustion of biogas containing siloxanes, silicon is released and can combine with free oxygen or other elements in the combustion gas. Deposits are formed containing mostly silica (SiO_2) or silicates (Si_xO_y) and can contain calcium, sulfur, zinc, phosphorus. Such *white mineral* deposits accumulate to a surface thickness of several millimeters and must be removed by chemical or mechanical means.

Practical and cost-effective technologies to remove siloxanes and other biogas contaminants are available.

For 1000 kg (wet weight) of input to a typical biodigester, total solids may be 30% of the wet weight while volatile suspended solids may be 90% of the total solids. Protein would be 20% of the volatile solids, carbohydrates would be 70% of the volatile solids, and finally fats would be 10% of the volatile solids.

Benefits of Manure Derived Biogas

High levels of methane are produced when manure is stored under anaerobic conditions. During storage and when manure has been applied to the land, nitrous oxide is also produced as a byproduct of the denitrification process. Nitrous oxide (N2O) is 320 times more aggressive as a greenhouse gas than carbon dioxide and methane 25 times more than carbon dioxide.

By converting cow manure into methane biogas via anaerobic digestion, the millions of cattle in the United States would be able to produce 100 billion kilowatt hours of electricity, enough to power millions of homes across the United States. In fact, one cow can produce enough manure in one day to generate 3 kilowatt hours of electricity; only 2.4 kilowatt hours of electricity are needed to power a single 100-watt light bulb for one day. Furthermore, by converting cattle manure into methane biogas instead of letting it decompose, global warming gases could be reduced by 99 million metric tons or 4%.

Applications

A biogas bus in Linköping, Sweden

Biogas can be used for electricity production on sewage works, in a CHP gas engine, where the waste heat from the engine is conveniently used for heating the digester; cooking; space heating; water heating; and process heating. If compressed, it can replace compressed natural gas for use in vehicles, where it can fuel an internal combustion engine or fuel cells and is a much more effective displacer of carbon dioxide than the normal use in on-site CHP plants.

Biogas Upgrading

Raw biogas produced from digestion is roughly 60% methane and 29% CO_2 with trace elements of H_2S; it is not of high enough quality to be used as fuel gas for machinery. The corrosive nature of H_2S alone is enough to destroy the internals of a plant.

Methane in biogas can be concentrated via a biogas upgrader to the same standards as fossil natural gas, which itself has to go through a cleaning process, and becomes *biomethane*. If the local gas network allows, the producer of the biogas may use their distribution networks. Gas must be very clean to reach pipeline quality and must be of the correct composition for the distribution network to accept. Carbon dioxide, water, hydrogen sulfide, and particulates must be removed if present.

There are four main methods of upgrading: water washing, pressure swing adsorption, selexol adsorption, and amine gas treating. In addition to these, the use of membrane separation technology for biogas upgrading is increasing, and there are already several plants operating in Europe and USA.

The most prevalent method is water washing where high pressure gas flows into a column where the carbon dioxide and other trace elements are scrubbed by cascading water running counter-flow to the gas. This arrangement could deliver 98% methane with manufacturers guaranteeing maximum 2% methane loss in the system. It takes roughly between 3% and 6% of the total energy output in gas to run a biogas upgrading system.

Biogas Gas-grid Injection

Gas-grid injection is the injection of biogas into the methane grid (natural gas grid). Injections includes biogas until the breakthrough of micro combined heat and power two-thirds of all the energy produced by biogas power plants was lost (the heat), using the grid to transport the gas to customers, the electricity and the heat can be used for on-site generation resulting in a reduction of losses in the transportation of energy. Typical energy losses in natural gas transmission systems range from 1% to 2%. The current energy losses on a large electrical system range from 5% to 8%.

Biogas in Transport

"Biogaståget Amanda" ("The Biogas Train Amanda") train near Linköping station, Sweden

If concentrated and compressed, it can be used in vehicle transportation. Compressed biogas is becoming widely used in Sweden, Switzerland, and Germany. A biogas-powered train, named Biogaståget Amanda (The Biogas Train Amanda), has been in service in Sweden since 2005. Biogas powers automobiles. In 1974, a British documentary film titled *Sweet as a Nut* detailed the biogas production process from pig manure and showed how it fueled a custom-adapted combustion engine. In 2007, an estimated 12,000 vehicles were being fueled with upgraded biogas worldwide, mostly in Europe.

Measuring in Biogas Environments

Biogas is part of the wet gas and condensing gas (or air) category that includes mist or fog in the gas stream. The mist or fog is predominately water vapor that condenses on the sides of pipes or stacks throughout the gas flow. Biogas environments include wastewater digesters, landfills, and animal feeding operations (covered livestock lagoons).

Ultrasonic flow meters are one of the few devices capable of measuring in a biogas atmosphere. Most thermal flow meters are unable to provide reliable data because the moisture causes steady high flow readings and continuous flow spiking, although there are single-point insertion thermal mass flow meters capable of accurately monitoring biogas flows with minimal pressure drop. They can handle moisture variations that occur in the flow stream because of daily and seasonal temperature fluctuations, and account for the moisture in the flow stream to produce a dry gas value.

Legislation

European Union

The European Union has legislation regarding waste management and landfill sites called the Landfill Directive.

Countries such as the United Kingdom and Germany now have legislation in force that provides farmers with long-term revenue and energy security.

United States

The United States legislates against landfill gas as it contains VOCs. The United States Clean Air Act and Title 40 of the Code of Federal Regulations (CFR) requires landfill owners to estimate the quantity of non-methane organic compounds (NMOCs) emitted. If the estimated NMOC emissions exceeds 50 tonnes per year, the landfill owner is required to collect the gas and treat it to remove the entrained NMOCs. Treatment of the landfill gas is usually by combustion. Because of the remoteness of landfill sites, it is sometimes not economically feasible to produce electricity from the gas.

Global Developments

United States

With the many benefits of biogas, it is starting to become a popular source of energy and is starting to be used in the United States more. In 2003, the United States consumed 147 trillion BTU of energy from "landfill gas", about 0.6% of the total U.S. natural gas consumption. Methane biogas derived from cow manure is being tested in the U.S. According to a 2008 study, collected by the *Science and Children* magazine, methane biogas from cow manure would be sufficient to produce 100 billion kilowatt hours enough to power millions of homes across America. Furthermore, methane biogas has been tested to prove that it can reduce 99 million metric tons of greenhouse gas emissions or about 4% of the greenhouse gases produced by the United States.

In Vermont, for example, biogas generated on dairy farms was included in the CVPS Cow Power program. The program was originally offered by Central Vermont Public Service Corporation as a voluntary tariff and now with a recent merger with Green Mountain Power is now the GMP Cow Power Program. Customers can elect to pay a premium on their electric bill, and that premium is passed directly to the farms in the program. In Sheldon, Vermont, Green Mountain Dairy has provided renewable energy as part of the Cow Power program. It started when the brothers who own the farm, Bill and Brian Rowell, wanted to address some of the manure management challenges faced by dairy farms, including manure odor, and nutrient availability for the crops they need to grow to feed the animals. They installed an anaerobic digester to process the cow and milking center waste from their 950 cows to produce renewable energy, a bedding to replace sawdust, and a plant-friendly fertilizer. The energy and environmental attributes are sold to the GMP Cow Power program. On average, the system run by the Rowells produces enough electricity to power 300 to 350 other homes. The generator capacity is about 300 kilowatts.

In Hereford, Texas, cow manure is being used to power an ethanol power plant. By switching to methane biogas, the ethanol power plant has saved 1000 barrels of oil a day. Over all, the power plant has reduced transportation costs and will be opening many more jobs for future power plants that will rely on biogas.

In Oakley, Kansas, an ethanol plant considered to be one of the largest biogas facilities in North America is using Integrated Manure Utilization System "IMUS" to produce heat for its boilers by utilizing feedlot manure, municipal organics and ethanol plant waste. At full capacity the plant is expected to replace 90% of the fossil fuel used in the manufacturing process of ethanol.

Europe

The level of development varies greatly in Europe. While countries such as Germany, Austria and Sweden are fairly advanced in their use of biogas, there is a vast potential for this renewable energy source in the rest of the continent, especially in Eastern Europe. Different legal frameworks, education schemes and the availability of technology are among the prime reasons behind this untapped potential. Another challenge for the further progression of biogas has been negative public perception.

Initiated by the events of the gas crisis in Europe during December 2008, it was decided to launch the EU project "SEBE" (Sustainable and Innovative European Biogas Environment) which is financed under the CENTRAL programme. The goal is to address the energy dependence of Europe by establishing an online platform to combine knowledge and launch pilot projects aimed at raising awareness among the public and developing new biogas technologies.

In February 2009, the European Biogas Association (EBA) was founded in Brussels as a non-profit organisation to promote the deployment of sustainable biogas production and use in Europe. EBA's strategy defines three priorities: establish biogas as an important part of Europe's energy mix, promote source separation of household waste to increase the gas potential, and support the production of biomethane as vehicle fuel. In July 2013, it had 60 members from 24 countries across Europe.

UK

As of September 2013, there are about 130 non-sewage biogas plants in the UK. Most are on-farm, and some larger facilities exist off-farm, which are taking food and consumer wastes.

On 5 October 2010, biogas was injected into the UK gas grid for the first time. Sewage from over 30,000 Oxfordshire homes is sent to Didcot sewage treatment works, where it is treated in an anaerobic digestor to produce biogas, which is then cleaned to provide gas for approximately 200 homes.

In 2015 the Green-Energy company Ecotricity announced their plans to build three grid-injecting digester's.

Germany

Germany is Europe's biggest biogas producer and the market leader in biogas technology. In 2010 there were 5,905 biogas plants operating throughout the country: Lower Saxony, Bavaria, and the eastern federal states are the main regions. Most of these plants are employed as power plants. Usually the biogas plants are directly connected with a CHP which produces electric power by burning the bio methane. The electrical power is then fed into the public power grid. In 2010, the total installed electrical capacity of these power plants was 2,291 MW. The electricity supply was approximately 12.8 TWh, which is 12.6% of the total generated renewable electricity.

Biogas in Germany is primarily extracted by the co-fermentation of energy crops (called 'NawaRo', an abbreviation of *nachwachsende Rohstoffe*, German for renewable resources) mixed with manure. The main crop used is corn. Organic waste and industrial and agricultural residues such as waste from the food industry are also used for biogas generation.In this respect, biogas production in Germany differs significantly from the UK, where biogas generated from landfill sites is most common.

Biogas production in Germany has developed rapidly over the last 20 years. The main reason is the legally created frameworks. Government support of renewable energy started in 1991 with the Electricity Feed-in Act (*StrEG*). This law guaranteed the producers of energy from renewable sources the feed into the public power grid, thus the power companies were forced to take all produced energy from independent private producers of green energy. In 2000 the Electricity Feed-in Act was replaced by the Renewable Energy Sources Act (*EEG*). This law even guaranteed a fixed compensation for the produced electric power over 20 years. The amount of around 8 ¢/kWh gave farmers the opportunity to become energy suppliers and gain a further source of income.

The German agricultural biogas production was given a further push in 2004 by implementing the so-called NawaRo-Bonus. This is a special payment given for the use of renewable resources, that is, energy crops. In 2007 the German government stressed its intention to invest further effort and support in improving the renewable energy supply to provide an answer on growing climate challenges and increasing oil prices by the 'Integrated Climate and Energy Programme'.

This continual trend of renewable energy promotion induces a number of challenges facing the management and organisation of renewable energy supply that has also several impacts on the biogas production. The first challenge to be noticed is the high area-consuming of the biogas electric power supply. In 2011 energy crops for biogas production consumed an area of circa 800,000 ha in Germany. This high demand of agricultural areas generates new competitions with the food industries that did not exist hitherto. Moreover, new industries and markets were created in predominately rural regions entailing different new players with an economic, political and civil background. Their influence and acting has to be governed to gain all advantages this new source of energy is offering. Finally biogas will furthermore play an important role in the German renewable energy supply if good governance is focused.

Indian Subcontinent

Biogas in India has been traditionally based on dairy manure as feed stock and these "gobar" gas plants have been in operation for a long period of time, especially in rural India. In the last 2-3 decades, research organisations with a focus on rural energy security have enhanced the design of the systems resulting in newer efficient low cost designs such as the Deenabandhu model.

The Deenabandhu Model is a new biogas-production model popular in India. (*Deenabandhu* means "friend of the helpless.") The unit usually has a capacity of 2 to 3 cubic metres. It is constructed using bricks or by a ferrocement mixture. In India, the brick model costs slightly more than the ferrocement model; however, India's Ministry of New and Renewable Energy offers some subsidy per model constructed.

LPG (Liquefied Petroleum Gas) is a key source of cooking fuel in urban India and its prices have been increasing along with the global fuel prices. Also the heavy subsidies provided by the successive governments in promoting LPG as a domestic cooking fuel has become a financial burden renewing the focus on biogas as a cooking fuel alternative in urban establishments. This has led to the development of prefabricated digester for modular deployments as compared to RCC and cement structures which take a longer duration to construct. Renewed focus on process technology like the Biourja process model has enhanced the stature of medium and large scale anaerobic digester in India as a potential alternative to LPG as primary cooking fuel.

In India, Nepal, Pakistan and Bangladesh biogas produced from the anaerobic digestion of manure in small-scale digestion facilities is called gobar gas; it is estimated that such facilities exist in over 2 million households in India, 50,000 in Bangladesh and thousands in Pakistan, particularly North Punjab, due to the thriving population of livestock. The digester is an airtight circular pit made of concrete with a pipe connection. The manure is directed to the pit, usually straight from the cattle shed. The pit is filled with a required quantity of wastewater. The gas pipe is connected to the kitchen fireplace through control valves. The combustion of this biogas has very little odour or smoke. Owing to simplicity in implementation and use of cheap raw materials in villages, it is one of the most environmentally sound energy sources for rural needs. One type of these system is the Sintex Digester. Some designs use vermiculture to further enhance the slurry produced by the biogas plant for use as compost.

To create awareness and associate the people interested in biogas, the Indian Biogas Association was formed. It aspires to be a unique blend of nationwide operators, manufacturers and planners of biogas plants, and representatives from science and research. The association was founded in 2010 and is now ready to start mushrooming. Its motto is "propagating Biogas in a sustainable way".

In Pakistan, the Rural Support Programmes Network is running the Pakistan Domestic Biogas Programme which has installed 5,360 biogas plants and has trained in excess of 200 masons on the technology and aims to develop the Biogas Sector in Pakistan.

In Nepal, the government provides subsidies to build biogas plant.

China

The Chinese had experimented the applications of biogas since 1958. Around 1970, China had installed 6,000,000 digesters in an effort to make agriculture more efficient. During the last years the technology has met high growth rates. This seems to be the earliest developments in generating biogas from agricultural waste.

In Developing Nations

Domestic biogas plants convert livestock manure and night soil into biogas and slurry, the fermented manure. This technology is feasible for small-holders with livestock producing 50 kg manure per day, an equivalent of about 6 pigs or 3 cows. This manure has to be collectable to mix it with water and feed it into the plant. Toilets can be connected. Another precondition is the temperature that affects the fermentation process. With an optimum at 36 C° the technology especially applies for those living in a (sub) tropical climate. This makes the technology for small holders in developing countries often suitable.

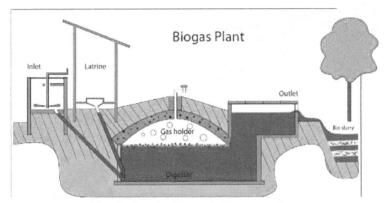

Simple sketch of household biogas plant

Depending on size and location, a typical brick made fixed dome biogas plant can be installed at the yard of a rural household with the investment between US$300 to $500 in Asian countries and up to $1400 in the African context. A high quality biogas plant needs minimum maintenance costs and can produce gas for at least 15–20 years without major problems and re-investments. For the user, biogas provides clean cooking energy, reduces indoor air pollution, and reduces the time needed for traditional biomass collection, especially for women and children. The slurry is a clean organic fertilizer that potentially increases agricultural productivity.

Domestic biogas technology is a proven and established technology in many parts of the world, especially Asia. Several countries in this region have embarked on large-scale programmes on domestic biogas, such as China and India.

The Netherlands Development Organisation, SNV, supports national programmes on domestic biogas that aim to establish commercial-viable domestic biogas

sectors in which local companies market, install and service biogas plants for households. In Asia, SNV is working in Nepal, Vietnam, Bangladesh, Bhutan, Cambodia, Lao PDR, Pakistan and Indonesia, and in Africa; Rwanda, Senegal, Burkina Faso, Ethiopia, Tanzania, Uganda, Kenya, Benin and Cameroon.

In South Africa a prebuilt Biogas system is manufactured and sold. One key feature is that installation requires less skill and is quicker to install as the digester tank is pre-made plastic.

Society and Culture

In the 1985 Australian film Mad Max Beyond Thunderdome the post-apocalyptic settlement Barter town is powered by a central biogas system based upon a piggery. As well as providing electricity, methane is used to power Barter's vehicles.

"Cow Town", written in the early 1940s, discuss the travails of a city vastly built on cow manure and the hardships brought upon by the resulting methane biogas. Carter McCormick, an engineer from a town outside the city, is sent in to figure out a way to utilize this gas to help power, rather than suffocate, the city.

Biochar

A piece of biochar.

Biochar is charcoal used as a soil amendment. Like most charcoal, biochar is made from biomass via pyrolysis. Biochar is under investigation as an approach to carbon sequestration to produce negative carbon dioxide emissions. Biochar thus has the potential to help mitigate climate change via carbon sequestration. Independently, biochar can increase soil fertility of acidic soils (low pH soils), increase agricultural productivity, and provide protection against some foliar and soil-borne diseases. Furthermore, biochar reduces pressure on forests. Biochar is a stable solid, rich in carbon, and can endure in soil for thousands of years.

History

Pre-Columbian Amazonians are believed to have used biochar to enhance soil productivity. They produced it by smoldering agricultural waste (i.e., covering burning biomass with soil) in pits or trenches. European settlers called it terra preta de Indio. Following observations and experiments, a research team working in French Guiana hypothesized that the Amazonian earthworm *Pontoscolex corethrurus* was the main agent of fine powdering and incorporation of charcoal debris to the mineral soil.

The term "biochar" was coined by Peter Read to describe charcoal used as a soil improvement.

Production

Biochar is a high-carbon, fine-grained residue that today is produced through modern pyrolysis processes, which is the direct thermal decomposition of biomass in the absence of oxygen, which prevents combustion, to obtain an array of solid (biochar), liquid (bio-oil), and gas (syngas) products. The specific yield from the pyrolysis is dependent on process conditions. such as temperature, and can be optimized to produce either energy or biochar. Temperatures of 400–500 °C (752–932 °F) produce more char, while temperatures above 700 °C (1,292 °F) favor the yield of liquid and gas fuel components. Pyrolysis occurs more quickly at the higher temperatures, typically requiring seconds instead of hours. High temperature pyrolysis is also known as gasification, and produces primarily syngas. Typical yields are 60% bio-oil, 20% biochar, and 20% syngas. By comparison, slow pyrolysis can produce substantially more char (~50%). Once initialized, both processes produce net energy. For typical inputs, the energy required to run a "fast" pyrolyzer is approximately 15% of the energy that it outputs. Modern pyrolysis plants can use the syngas created by the pyrolysis process and output 3–9 times the amount of energy required to run.

The Amazonian pit/trench method harvests neither bio-oil nor syngas, and releases a large amount of CO_2, black carbon, and other greenhouse gases (GHG)s (and potentially, toxins) into the air. Commercial-scale systems process agricultural waste, paper byproducts, and even municipal waste and typically eliminate these side effects by capturing and using the liquid and gas products.

Centralized, Decentralized, and Mobile Systems

In a centralized system, all biomass in a region is brought to a central plant for processing. Alternatively, each farmer or group of farmers can operate a lower-tech kiln. Finally, a truck equipped with a pyrolyzer can move from place to place to pyrolyze biomass. Vehicle power comes from the syngas stream, while the biochar remains on the farm. The biofuel is sent to a refinery or storage site. Factors that influence the choice of system type include the cost of transportation of the liquid and solid byproducts, the amount of material to be processed, and the ability to feed directly into the power grid.

For crops that are not exclusively for biochar production, the residue-to-product ratio (RPR) and the collection factor (CF) the percent of the residue not used for other things, measure the approximate amount of feedstock that can be obtained for pyrolysis after harvesting the primary product. For instance, Brazil harvests approximately 460 million tons (MT) of sugarcane annually, with an RPR of 0.30, and a CF of 0.70 for the sugarcane tops, which normally are burned in the field. This translates into approximately 100 MT of residue annually, which could be pyrolyzed to create energy and soil additives. Adding in the bagasse (sugarcane waste) (RPR=0.29 CF=1.0), which is otherwise burned (inefficiently) in boilers, raises the total to 230 MT of pyrolysis feedstock. Some plant residue, however, must remain on the soil to avoid increased costs and emissions from nitrogen fertilizers.

Pyrolysis technologies for processing loose and leafy biomass produce both biochar and syngas.

Thermo-catalytic Depolymerization

Alternatively, "thermo-catalytic depolymerization", which utilizes microwaves, has recently been used to efficiently convert organic matter to biochar on an industrial scale, producing ~50% char.

Uses

Carbon Sink

The burning and natural decomposition of biomass and in particular agricultural waste adds large amounts of $CO2$ to the atmosphere. Biochar that is stable, fixed, and 'recalcitrant' carbon can store large amounts of greenhouse gases in the ground for centuries, potentially reducing or stalling the growth in atmospheric greenhouse gas levels; at the same time its presence in the earth can improve water quality, increase soil fertility, raise agricultural productivity, and reduce pressure on old-growth forests.

Biochar can sequester carbon in the soil for hundreds to thousands of years, like coal. Such a carbon-negative technology would lead to a net withdrawal of CO_2 from the atmosphere, while producing and consuming energy. This technique is advocated by prominent scientists such as James Hansen, head of the NASA Goddard Institute for Space Studies, and James Lovelock, creator of the Gaia hypothesis, for mitigation of global warming by greenhouse gas remediation.

Researchers have estimated that sustainable use of biocharring could reduce the global net emissions of carbon dioxide ($CO2$), methane, and nitrous oxide by up to 1.8 Pg $CO2$-C equivalent ($CO2$-C_e) per year (12% of current anthropogenic $CO2$-Ce emissions; 1 Pg=1 Gt), and total net emissions over the course of the next century by 130 Pg $CO2$-C_e, without endangering food security, habitat, or soil conservation.

Soil Amendment

Biochar is recognised as offering a number of benefits for soil health. Many benefits are related to the extremely porous nature of biochar. This structure is found to be very effective at retaining both water and water-soluble nutrients. Soil biologist Elaine Ingham indicates the extreme suitability of biochar as a habitat for many beneficial soil micro organisms. She points out that when pre charged with these beneficial organisms biochar becomes an extremely effective soil amendment promoting good soil, and in turn plant, health.

Biochar has also been shown to reduce leaching of E-coli through sandy soils depending on application rate, feedstock, pyrolysis temperature, soil moisture content, soil texture, and surface properties of the bacteria.

For plants that require high potash and elevated pH, biochar can be used as a soil amendment to improve yield.

Biochar can improve water quality, reduce soil emissions of greenhouse gases, reduce nutrient leaching, reduce soil acidity, and reduce irrigation and fertilizer requirements. Biochar was also found under certain circumstances to induce plant systemic responses to foliar fungal diseases and to improve plant responses to diseases caused by soilborne pathogens.

The various impacts of biochar can be dependent on the properties of the biochar, as well as the amount applied, and there is still a lack of knowledge about the important mechanisms and properties. Biochar impact may depend on regional conditions including soil type, soil condition (depleted or healthy), temperature, and humidity. Modest additions of biochar to soil reduce nitrous oxide N2O emissions by up to 80% and eliminate methane emissions, which are both more potent greenhouse gases than CO_2.

Studies have reported positive effects from biochar on crop production in degraded and nutrient–poor soils. Biochar can be designed with specific qualities to target distinct properties of soils. Biochar reduces leaching of critical nutrients, creates a higher crop uptake of nutrients, and provides greater soil availability of nutrients. At 10% levels biochar reduced contaminant levels in plants by up to 80%, while reducing total chlordane and DDX content in the plants by 68 and 79%, respectively. On the other hand, because of its high adsorption capacity, biochar may reduce the efficacy of soil applied pesticides that are needed for weed and pest control. High-surface-area biochars may be particularly problematic in this regard; more research into the long-term effects of biochar addition to soil is needed.

Slash-and-char

Switching from slash-and-burn to slash-and-char farming techniques in Brazil can decrease both deforestation of the Amazon basin and carbon dioxide emission, as well as increase crop yields. Slash-and-burn leaves only 3% of the carbon from the organic material in the soil.

Slash-and-char can keep up to 50% of the carbon in a highly stable form. Returning the biochar into the soil rather than removing it all for energy production reduces the need for nitrogen fertilizers, thereby reducing cost and emissions from fertilizer production and transport. Additionally, by improving the soil's ability to be tilled, fertility, and productivity, biochar–enhanced soils can indefinitely sustain agricultural production, whereas non-enriched soils quickly become depleted of nutrients, forcing farmers to abandon the fields, producing a continuous slash and burn cycle and the continued loss of tropical rainforest. Using pyrolysis to produce bio-energy also has the added benefit of not requiring infrastructure changes the way processing biomass for cellulosic ethanol does. Additionally, the biochar produced can be applied by the currently used machinery for tilling the soil or equipment used to apply fertilizer.

Water Retention

Biochar is a desirable soil material in many locations due to its ability to attract and retain water. This is possible because of its porous structure and high surface area. As a result, nutrients, phosphorus, and agrochemicals are retained for the plants benefit. Plants therefore, are healthier and fertilizers leach less into surface or groundwater.

Energy Production: Bio-oil and Syngas

Mobile pyrolysis units can be used to lower the costs of transportation of the biomass if the biochar is returned to the soil and the syngas stream is used to power the process. Bio-oil contains organic acids that are corrosive to steel containers, has a high water vapor content that is detrimental to ignition, and, unless carefully cleaned, contains some biochar particles which can block injectors.

If biochar is used for the production of energy rather than as a soil amendment, it can be directly substituted for any application that uses coal. Pyrolysis also may be the most cost-effective way of electricity generation from biomaterial.

Direct and Indirect Benefits

- The pyrolysis of forest- or agriculture-derived biomass residue generates a bio-fuel without competition with crop production.

- Biochar is a pyrolysis byproduct that may be ploughed into soils in crop fields to enhance their fertility and stability, and for medium- to long-term carbon sequestration in these soils.

- Biochar enhances the natural process: the biosphere captures CO_2, especially through plant production, but only a small portion is stably sequestered for a relatively long time (soil, wood, etc.).

- Biomass production to obtain biofuels and biochar for carbon sequestration in the soil is a carbon-negative process, i.e. more CO2 is removed from the atmosphere than released, thus enabling long-term sequestration.

Research

Intensive research into manifold aspects involving the pyrolysis/biochar platform is underway around the world. From 2005 to 2012, there were 1,038 articles that included the word "biochar" or "bio-char" in the topic that had been indexed in the ISI Web of Science. Further research is in progress by such diverse institutions around the world as Cornell University, the University of Edinburgh, which has a dedicated research unit., and the Agricultural Research Organization (ARO) of Israel, Volcani Center, where a network of researchers involved in biochar research (iBRN, Israel Biochar Researchers Network) was established as early as 2009.

Students at Stevens Institute of Technology in New Jersey are developing supercapacitors that use electrodes made of biochar. A process developed by University of Florida researchers that removes phosphate from water, also yields methane gas usable as fuel and phosphate-laden carbon suitable for enriching soil.

Emerging Commercial Sector

Calculations suggest that emissions reductions can be 12 to 84% greater if biochar is put back into the soil instead of being burned to offset fossil-fuel use. Thus biochar sequestration offers the chance to turn bioenergy into a carbon-negative industry.

Johannes Lehmann, of Cornell University, estimates that pyrolysis can be cost-effective for a combination of sequestration and energy production when the cost of a CO_2 ton reaches $37. As of mid-February 2010, CO_2 is trading at $16.82/ton on the European Climate Exchange (ECX), so using pyrolysis for bioenergy production may be feasible even if it is more expensive than fossil fuel.

Current biochar projects make no significant impact on the overall global carbon budget, although expansion of this technique has been advocated as a geoengineering approach. In May 2009, the Biochar Fund received a grant from the Congo Basin Forest Fund for a project in Central Africa to simultaneously slow down deforestation, increase the food security of rural communities, provide renewable energy and sequester carbon.

Application rates of 2.5–20 tonnes per hectare (1.0–8.1 t/acre) appear to be required to produce significant improvements in plant yields. Biochar costs in developed countries vary from $300–7000/tonne, generally too high for the farmer/horticulturalist and prohibitive for low-input field crops. In developing countries, constraints on agricultural biochar relate more to biomass availability and production time. An alternative is to use small amounts of biochar in lower cost biochar-fertilizer complexes.

Various companies in North America, Australia, and England sell biochar or biochar production units. In England Carbon Gold supply a range of biochar-based soil improvers, composts and fertilisers for arboriculture, horticulture and turfcare as well as to home growers. In Sweden the 'Stockholm Solution' is an urban tree planting system that uses 30% biochar to support healthy growth of the urban forest. The Qatar Aspire Park now uses biochar to help trees cope with the intense heat of their summers.

At the 2009 International Biochar Conference, a mobile pyrolysis unit with a specified intake of 1,000 pounds (450 kg) was introduced for agricultural applications. The unit had a length of 12 feet and height of 7 feet (3.6 m by 2.1m).

A production unit in Dunlap, Tennessee by Mantria Corporation opened in August 2009 after testing and an initial run, was later shut down as part of a Ponzi scheme investigation.

Bioliquids

Bioliquids are liquid fuels made from biomass for energy purposes other than transport (i.e. heating and electricity).

Bioliquids are usually made from virgin or used vegetable and seed oils, like palm or soya oil. These oils are burned in a power station to create heat, which can then be used to warm homes or boil water to make steam. This steam can then be used to drive a turbine to generate electricity.

Rudolf Diesel's first public exhibition of the internal combustion engine, that was to later bear his name, ran on peanut oil.

Bioliquid Production and Use

Bioliquids have been used for many years to provide heat for homes on a small scale but now big energy providers are looking at their use on a larger scale.

A controversial plant in Bristol (UK) was recently given the go ahead despite receiving several hundred complaints. The plant will be built and operated by W4B and provide enough power for 25,000 homes.

Advantages

Bioliquids have several key advantages over other sources of renewable energy:

- Bioliquids have a high energy density
- The technology is well established, having been used for many years

- Can be used on demand, reacting quickly to changes in demand for power

- Can help reduce dependency on foreign oil.

- Reduces the green house gas emissions.

Disadvantages

Many of the same problems that affect biofuels also affect bioliquids and there are various social, economic, environmental and technical issues, which have been discussed in the popular media and scientific journals. These include: the effect of moderating oil prices, the "food vs fuel" debate, poverty reduction potential, carbon emissions levels, sustainable biofuel production, deforestation and soil erosion, loss of biodiversity, impact on water resources, as well as energy balance and efficiency.

Bioliquids also have several key problems compared to other sources of renewable energy:

- Price of fuel is very variable, due to competitiveness of feedstock for other uses (e.g. soap)

- Supply chain is still very new

- Governments, such as the EU, remained undecided on bioliquids

Blue Carbon

Blue carbon is the carbon captured by the world's oceans and coastal ecosystems. The carbon captured by living organisms in oceans is stored in the form of biomass and sediments from mangroves, salt marshes, seagrasses and potentially algae.

Overview

Historically the ocean and terrestrial forest ecosystems have been the major natural carbon (C) sinks. New research on the role of vegetated coastal ecosystems have highlighted their potential as highly efficient C sinks, and led to the scientific recognition of the term "Blue Carbon". "Blue Carbon" designates carbon that is fixed via ocean and coastal ecosystems, rather than traditional land ecosystems, like forests. Although the ocean's vegetated habitats cover less than 0.5% of the seabed, they are responsible for more than 50%, and potentially up to 70%, of all carbon storage in ocean sediments. Mangroves, salt marshes and seagrasses make up the majority of the ocean's vegetated habitats but only equal 0.05% of the plant biomass on land. Despite their small footprint, they can store a comparable amount of carbon per year and are highly efficient carbon sinks. Seagrasses, mangroves and salt marshes can capture carbon dioxide (CO_2) from the atmosphere by sequestering the C in their underlying sediments, in underground and below-ground biomass, and in dead biomass. In plant biomass such as leaves, stems, branches or roots, blue carbon can be sequestered for years to decades, and for thousands to millions of years in underlying plant sediments. Current estimates of long-term blue carbon C burial capacity are variable, and research is ongoing. Although vegetated coastal ecosystems cover less area and have less aboveground biomass than terrestrial plants they have the potential to impact longterm C sequestration, particularly in sediment sinks. One of the main concerns with Blue Carbon is the rate of loss of these important marine ecosystems is much higher than any other ecosystem on the planet, even compared to rainforests. Current estimates suggest a loss of 2-7% per year, which is not only lost carbon sequestration, but also lost habitat that is important for managing climate, coastal protection, and health.

Types of Blue Carbon Ecosystems

Seagrass

Seagrass at Rapid Bay Jetty, South Australia

Seagrass are a group of about 60 angiosperm species that have adapted to an aquatic life, and can grow in meadows along the shores of all continents except Antarctica. Seagrass meadows form in maximum depths of up to 50m, depending on water quality and light availability, and can include up to 12 different species in one meadow. These seagrass meadows are highly productive habitats that provide many ecosystem services, including sediment stabilization, habitat and biodiversity, better water quality, and carbon and nutrient sequestration. The current documented seagrass area is 177,000 km², but is thought to underestimate the total area since many areas with large seagrass meadows have not been thoroughly documented. Most common estimates are 300,000 to 600,000 km², with up to 4,320,000 km² suitable seagrass habitat worldwide. Although seagrass makes up only 0.1% of the area of the ocean floor, it accounts for approximately 10-18% of the total oceanic carbon burial. Currently global seagrass meadows are estimated to store as much as 19.9 Pg (gigaton, or billion tons) of organic carbon. Carbon primarily accumulates in marine sediments, which are anoxic and thus continually preserve organic carbon from decadal-millennial time scales. High accumulation rates, low oxygen, low sediment conductivity and slower microbial decomposition rates all encourage carbon burial and carbon accumulation in these coastal sediments. Compared to terrestrial habitats that lose carbon stocks as CO_2 during decomposition or by disturbances like fires or deforestation, marine carbon sinks can retain C for much longer time periods. Carbon sequestration rates in seagrass meadows vary depending on the species, characteristics of the sediment, and depth of the habitats, but on average the carbon burial rate is approximately 138 g C m^{-2} yr^{-1}. Seagrass habitats are threatened by coastal eutrophication, increased seawater temperatures, increased sedimentation and coastal development, and sea-level rise which may decrease light availability for photosynthesis. Seagrass loss has accelerated over the past few decades, from 0.9% per year prior to 1940 to 7% per year in 1990, with about 1/3 of global loss since WWII. Scientists encourage protection and continued research of these ecosystems for organic carbon storage, valuable habitat and other ecosystem services.

Mangrove

Mangrove forest in Everglades National Park, FL

Mangroves are woody halophytes that form intertidal forests and provide many important ecosystem services including coastal protection, nursery grounds for coastal fish and crustaceans, forest products, recreation, nutrient filtration and carbon sequestration. Currently they are found in 123 countries, with 73 identified species. They grow along coastlines in subtropical and tropical waters, depending mainly on temperature, but also vary with precipitation, tides, waves and water flow. Because they grow at the intersection between land and sea, they have semi-terrestrial and marine components, including unique adaptations including aerial roots, viviparous embryos, and highly efficient nutrient retention mechanisms. Mangroves cover approximately 150,000 km^2 worldwide, but have declined by 20% in the last 25 years, mainly due to coastal development and land conversion. Mangrove deforestation is slowing, from 1.04% loss per year in the 1980s to 0.66% loss in the early 2000s, as research and understanding of mangrove benefits have increased. Mangrove forests are responsible for approximately 10% of global carbon burial, with an estimated carbon burial rate of 174 g C m^{-2} yr^{-1}. Mangroves, like seagrasses, have potential for high levels of carbon sequestration. They account for 3% of the global carbon sequestration by tropical forests and 14% of the global coastal ocean's carbon burial. Mangroves are naturally disturbed by floods, tsunamis, coastal storms like cyclones and hurricanes, lightning, disease and pests, and changes in water quality or temperature. Although they are resilient to many of these natural disturbances, they are highly susceptible to human impacts including urban development, aquaculture, mining, and overexploitation of shellfish, crustaceans, fish and timber. Mangroves provide globally important ecosystem services and carbon sequestration and are thus an important habitat to conserve and repair when possible.

Marsh

Tidal marsh in Hilton Head, SC

Marshes, intertidal ecosystems dominated by herbaceous vegetation, can be found globally on coastlines from the arctic to the subtropics. In the tropics, marshes are replaced by mangroves as the dominant coastal vegetation. Marshes have high productivity, with a large portion of primary production in belowground biomass. This belowground biomass can form deposits up to 8m deep. Marshes provide valuable habitat for plants, birds, and juvenile fish, protect coastal habitat from storm surge and flooding, and can reduce nutrient loading to coastal waters. Similarly to mangrove and seagrass habitats, marshes also serve as important carbon sinks. Marshes sequester C in underground biomass due to high rates of organic sedimentation and anaerobic-dominated decomposition. Salt marshes cover approximately 22,000 to 400,000 km^2 globally, with an estimated carbon burial rate of 210 g C m^{-2} yr^{-1}. Tidal marshes have been impacted by humans for centuries, including modification for grazing, haymaking, reclamation for agriculture, development and ports, evaporation ponds for salt production, modification for aquaculture, insect control, tidal power and flood protection. Marshes are also susceptible to pollution from oil, industrial chemicals, and most commonly, eutrophication. Introduced species, sea-level rise, river damming and decreased sedimentation are additional longterm changes that affect marsh habitat, and in turn, may affect carbon sequestration potential.

Algae

Both macroalgae and microalgae are being investigated as possible means of carbon sequestration. Because algae lack the complex lignin associated with terrestrial plants, the carbon in algae is released into the atmosphere more rapidly than carbon captured on land. Algae have been proposed as a short-term storage pool of carbon that can be used as a feedstock for the production of various biogenic fuels. Microalgae are often put forth as a potential feedstock for carbon-neutral biodiesel and biomethane production due to their high lipid content. Macroalgae, on the other hand, do not have high lipid content and have limited potential as biodiesel feedstock, although they can still be used as feedstock for other biofuel generation. Macroalgae have also been investigated as a feedstock for the production of biochar. The biochar produced from macroalgae is higher in agriculturally important nutrients than biochar produced from terrestrial sources. Another novel approach to carbon capture which utilizes algae is the Bicarbonate-based Integrated Carbon Capture and Algae Production Systems (BICCAPS) developed by a collaboration between Washington State University in the United States and Dalian Ocean University in China. Many cyanobacteria, microalgae, and macroalgae species can utilize carbonate as a carbon source for photosynthesis. In the BICCAPS, alkaliphilic microalgae utilize carbon captured from flue gases in the form of bicarbonate. In South Korea, macroalgae have been utilized as part of a climate change mitigation program. The country has established the Coastal CO_2 Removal Belt (CCRB) which is composed of artificial and natural ecosystems. The goal is to capture carbon using large areas of kelp forest.

Ecosystem Restoration

Restoration of mangrove forests, seagrass meadows, marshes, and kelp forests has been implemented in many countries. These restored ecosystems have the potential to act as carbon sinks. Restored seagrass meadows were found to start sequestering carbon in sediment within about four years. This was the time needed for the meadow to reach sufficient shoot density to cause sediment deposition. Similarly, mangrove plantations in China showed higher sedimentation rates than barren land and lower sedimentation rates than established mangrove forests. This pattern in sedimentation rate is thought to be a function of the plantation's young age and lower vegetation density.

Nutrient Stoichiometry of Seagrasses

The primary nutrients determining sea grass growth are carbon (C), nitrogen (N), phosphorus (P), and light for photosynthesis. Nitrogen and P can be acquired from sediment pore water or from the water column, and sea grasses can uptake N in both ammonium (NH_4^+) and nitrate (NO_3^-) form.

A number of studies from around the world have found that there is a wide range in the concentrations of C, N, and P in seagrasses depending on their species and environmental factors. For instance, plants collected from high-nutrient environments had lower C:N and C:P ratios than plants collected from low-nutrient environments. Sea grass stoichiometry does not follow the Redfield ratio commonly used as an indicator of nutrient availability for phytoplankton growth. In fact, a number of studies from around the world have found that the proportion of C:N:P in sea grasses can vary significantly depending on their species, nutrient availability, or other environmental factors. Depending on environmental conditions, sea grasses can be either P-limited or N-limited.

An early study of sea grass stoichiometry suggested that the "Redfield" balanced ratio between N and P for sea grasses is approximately 30:1. However, N and P concentrations are strictly not correlated, suggesting that sea grasses can adapt their nutrient uptake based on what is available in the environment. For example, sea grasses from meadows fertilized with bird excrement have shown a higher proportion of phosphate than unfertilized meadows. Alternately, sea grasses in environments with higher loading rates and organic matter diagenesis supply more P, leading to N-limitation. P availability in *T. testudinum* is the limiting nutrient. The nutrient distribution in *T. testudinum* ranges from 29.4-43.3% C, 0.88-3.96% N, and 0.048-0.243% P. This equates to a mean ratio of 24.6 C:N, 937.4 C:P, and 40.2 N:P. This information can also be used to characterize the nutrient availability of a bay or other water body (which is difficult to measure directly) by sampling the sea grasses living there.

Light availability is another factor that can affect the nutrient stoichiometry of sea grasses. Nutrient limitation can only occur when photosynthetic energy causes grasses to grow faster than the influx of new nutrients. For example, low light environments tend to have a lower C:N ratio. Alternately, high-N environments can have an indirect negative effect to sea grass growth by promoting growth of algae that reduce the total amount of available light.

Nutrient variability in sea grasses can have potential implications for wastewater management in coastal environments. High amounts of anthropogenic nitrogen discharge could cause eutrophication in previously N-limited environments, leading to hypoxic conditions in the sea grass meadow and affecting the carrying capacity of that ecosystem.

A study of annual deposition of C, N, and P from P. Oceanica sea grass meadows in northeast Spain found that the meadow sequestered 198 g C m-2 yr-1, 13.4 g N m-2 yr-1, and 2.01 g P m-2 yr-1 into the sediment. Subsequent remineralization of carbon from the sediments due to respiration returned approximately 8% of the sequestered carbon, or 15.6 g C m-2 yr -1.

Distribution and Decline of Blue Carbon Ecosystems

Seagrasses, mangroves and marshes are types of vegetated coastal habitats that cover approximately 49 million hectares worldwide. Seagrass ecosystems range from polar to tropical regions, mangroves are found in tropical and sub-tropical ecosystems and tidal marshes are found in mostly temperate regions such as on the east coast of the United States. As habitats that sequester carbon are altered and decreased, that stored amount of C is being released into the atmosphere, continuing the current accelerated rate of climate change. Impacts on these habitats globally will directly and indirectly release the previously stored carbon, which had been sequestered in sediments of these habitats. Declines of vegetated coastal habitats are seen worldwide; examples seen in mangroves are due to clearing for shrimp ponds such is the case in Indonesia, while in seagrasses there are both natural causes due to pathogens and may be exacerbated by anthropogenic effects. Quantifying rates of decrease are difficult to calculate, however measurements have been estimated by researchers indicating that if blue carbon ecosystems continue to decline, for any number of reasons, 30-40% of tidal marshes and seagrasses and approximately 100% of mangroves could be gone in the next century.

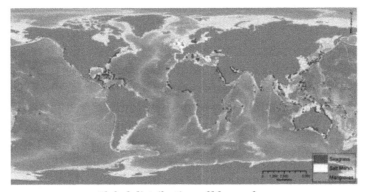

Global distribution of blue carbon

Decline in seagrasses are due to a number of factors including drought, water quality issues, agricultural practices, invasive species, pathogens, fishing and climate change. Over 35% of global mangrove habitat remains. Decreases in habitat is due to damming of rivers, clearing for aquaculture, development etc., overfishing, and climate change, according to the World Wildlife Fund. Nearly 16% of mangroves assessed by the IUCN are on the IUCN Red List; due to development and other causes 1 in 6 worldwide mangroves are in threat of extinction. Dams threaten habitats by slowing the amount of freshwater reaching mangroves. Coral reef destruction also plays a role in mangrove habitat health as reefs slow wave energy to a level that mangroves are more tolerant of. Salt marshes may not be expansive worldwide in relation to forests, but they have a C burial rate that is over 50 times faster than tropical rainforests. Rates of burial have been estimated at up to 87.2 ± 9.6 Tg C yr-1 which is greater than that of tropical rainforests, 53 ± 9.6 Tg C yr-1. Since the 1800s salt marshes have been disturbed due to development and a lack of understanding their importance. The 25% decline since that time has led to a decrease in potential C sink area coupled with the release of once buried C. Consequences of increasingly degraded marsh habitat are a decrease in C stock in sediments, a decrease in plant biomass and thus a decrease in photosynthesis reducing the amount of CO_2 taken up by the plants, failure of C in plant blades to be transferred into the sediment, possible acceleration of erosive processes due to the lack of plant biomass, and acceleration of buried C release to the atmosphere.

Reasons for decline of mangroves, seagrass, and marshes include land use changes, climate and drought related effects, dams built in the watershed, convergence to aquaculture and agriculture, land development and sea-level rise due to climate change. Increases in these activities can lead to significant decreases in habitat available and thus increases in released C from sediments. As anthropogenic effects and climate change are heightened, the effectiveness of blue carbon sinks will diminish and CO_2 emissions will be further increased. Data on the rates at which CO_2 is being released into the atmosphere is not robust currently, however research is being conducted to gather a better information to analyze trends. Loss of underground biomass (roots and rhizomes) will allow for CO_2 to be emitted changing these habitats into sources rather than carbon sinks.

Sedimentation and Blue Carbon Burial

Organic carbon is only sequestered from the oceanic system if it reaches the sea floor and gets covered by a layer of sediment. Reduced oxygen levels in buried environments mean that tiny bacteria who eat organic matter and respire CO_2 can't decompose the carbon, so it is removed from the system permanently. Organic matter that sinks but is not buried by a sufficiently deep layer of sediment is subject to re-suspension by changing ocean currents, bioturbation by organisms that live in the top layer of marine sediments, and decomposition by heterotrophic bacteria. If any of these processes occur, the organic carbon is released back into the system. Carbon sequestration takes place only if burial rates by sediment are greater than the long term rates of erosion, bioturbation, and decomposition.

Spatial Variability in Sedimentation

Sedimentation is the rate at which floating or suspended particulate matter sinks and accumulates on the ocean floor. The faster (more energetic) the current, the more sediment it can pick up. As sediment laden currents slow, the particles fall out of suspension and come to rest on the sea floor. In other words, a fast current can pick up lots of heavy grains, where as a slow current can pick up only tiny pieces. As one can imagine, different places in the ocean vary drastically when it comes to the amount of suspended sediment and rate of deposition.

Open Ocean

The open ocean has very low sedimentation rates because most of the sediments that make it here are carried by the wind. Wind transport accounts for only a small fraction of the total sediment delivery to the oceans. Additionally, there is much less plant and animal life living in the open ocean that could be buried. Therefore, carbon burial rates are relatively slow in the open ocean.

Coastal Margins

Coastal margins have high sedimentation rates due to sediment input by rivers, which account for the vast majority of sediment delivery to the ocean. In most cases, sediments are deposited near the river mouth or are transported in the alongshore direction due to wave forcing. In some places sediment falls into submarine canyons and is transported off-shelf, if the canyon is sufficiently large or the shelf is narrow. Coastal margins also contain diverse and plentiful marine species, especially in paces that experience periodic upwelling. More marine life combined with higher sedimentation rates on coastal margins creates hotspots for carbon burial.

Submarine Canyons

Marine canyons are magnets for sediment because as currents carry sediment on the shelf in the alongshore direction, the path of the current crosses canyons perpendicularly. When the same amount of water flow is suddenly in much deeper water it slows down and deposits sediment. Due to the extreme depositional environment, carbon burial rates in the Nazare Canyon near Portugal are 30 times greater than the adjacent continental slope! This canyon alone accounts for about 0.03% of global terrestrial organic carbon burial in marine sediments. This may not seem like much, but the Nazarre submarine canyon only makes up 0.0001% of the area of the worlds ocean floor.

Human Changes to Global Sedimentary Systems

Humans have been modifying sediment cycles on a massive scale for thousands of years through a number of mechanisms.

Agriculture/Land Clearing

The first major change to global sedimentary cycling happened when humans started clearing land to grow crops. In a natural ecosystem, roots from plants hold sediment in place when it rains. Trees and shrubs reduce the amount of rainfall that impacts the dirt, and create obstacles that forest streams must flow around. When all vegetation is removed rainfall impacts directly on the dirt, there are no roots to hold on to the sediment, and there is nothing to stop the stream from scouring banks as it flows straight downhill. Because of this, land clearing causes an increase in erosion rates when compared to a natural system.

Dams

The first dams date back to 3000 BC and were built to control flood waters for agriculture. When sediment laden river flow reaches a dam's reservoir, the water slows down as it pools. Since slower water can't carry as much sediment, virtually all of the sediment falls out of suspension before the water passes through the dam. The result is that most dams are nearly 100% efficient sediment traps. Additionally, the use of dams for flood control reduces the ability of downstream channels to produce sediment. Since the vast majority of sedimentation occurs during the biggest floods, reduced frequency and intensity of flood-like flows can drastically change production rates. For thousands of years there were too few dams to have a significant impact on global sedimentary cycles, except for local impacts on a few river deltas such as the Nile which were significant. However The popularization of hydroelectric power in the last century has caused an enormous boom in dam building. Currently only a third of the world's largest rivers flow unimpeded to the ocean.

Channelization

In a natural system, the banks of a river will meander back and forth as different channels erode, accrete, open, or close. Seasonal floods regularly overwhelm riverbanks and deposit nutrients on adjacent flood plains. These services are essential to natural ecosystems, but can be troublesome for humans, who love to build infrastructure and development close to rivers. In response, rivers in populated areas are often channelized, meaning that their banks and sometimes beds are armored with a hard material, such as rocks or concrete, which prevent erosion and fixes the stream in place. This inhibits sedimentation because there is much less soft substrate left for the river to take downstream.

Impacts

Currently, the net effect of humans on global sedimentary cycling is a drastic reduction in the amount of sediment that makes it to the ocean. If we continue to build dams and channelize rivers, we will continue to see a number of problems in coastal areas including sinking deltas, shrinking beaches, and disappearing salt marshes. In addition, it's possible that we might ruin the ability of coastal margins to bury blue carbon. Without sequestration of carbon in coastal marine sediments, we will likely see accelerated global climate change.

Other Factors Influencing Blue Carbon Burial Rates

Density of Vegetation

The density of vegetation in mangrove forests, seagrass meadows, and tidal marshes is an important factor in carbon burial rates.The density of the vegetation must be sufficient to change water flows enough to reduce erosion and increase sediment deposition.

Nutrient Load

Increases in carbon capture and sequestration have been observed in both mangrove and seagrass ecosystems which have been subjected to high nutrient loads, either intentionally or due to waste from human activities. Intentional fertilization has been used in seagrass meadow restoration. Perches for seabirds are installed in the meadow and the bird droppings are the fertilizer source. The fertilization allows fast growing varieties of seagrasses to establish and grow. The species composition of these meadows is markedly different than the original seagrass meadow, although after the meadow has been reestablished and fertilization terminated, the meadows return to a species composition that more closely resembles an undisturbed meadow. Research done on mangrove soils from the Red Sea have shown that increases in nutrient loads to these soils do not increase carbon mineralization and subsequent CO_2 release. This neutral effect of fertilization was not found to be true in all mangrove forest types. Carbon capture rates also increased in these forests due to increased growth rates of the mangroves. In forests with increases in respiration there were also increases in mangrove growth of up to six times the normal rate.

Engineered Approaches to Blue Carbon

A US Department of Energy study from 2001 proposed to replicate a natural process of carbon sequestration in the ocean by combining water rich in CO_2 gas with carbonate $[CO_3^-]$ to produce a bicarbonate $[HCO_3^-]$ slurry. Practically, the engineered process could involve hydrating the CO_2 from power plant flue gas and running it through a porous bed of limestone to 'fix' the carbon in a saturated bicarbonate solution. This solution could then be deposited at sea to sink in the deep ocean. The cost of this process, from capture to ocean burial, was estimated to range between $90 to $180 per tonne of CO_2 and was highly dependent on the distance required to transport limestone, seawater, and the resulting bicarbonate solution.

Expected benefits from bicarbonate production over direct CO_2 gas injection would be a significantly smaller change in ocean acidity and a longer timescale for burial before the captured carbon would be released back to the atmosphere.

Cellulosic Sugars

Cellulosic sugars are derived from non-food biomass (e.g. wood, agricultural residues, municipal solid waste). The biomass is primarily composed of carbohydrate polymers cellulose, hemicellulose, and an aromatic polymer (lignin). The hemicellulose is a polymer of mainly five-carbon sugars $C_5H_{10}O_5$ (xylose). and the cellulose is a polymer of six-carbon sugar $C_6H_{12}O_6$ (glucose). Cellulose fibers are considered to be a plant's structural building blocks and are tightly bound to lignin, but the biomass can be deconstructed using Acid hydrolysis, enzymatic hydrolysis, organosolv dissolution, autohydrolysis or supercritical hydrolysis.

Biomass (cellulose, hemicellulose and lignocellulose) contain vast amounts of fermentable sugars. These sugars may be produced from a wide variety of feedstocks and can be converted into a multitude of biochemical, biofuel, and polymer products by either biological or chemical routes.

Industrial Use

In January 2012, BASF invested in Pennsylvania-based Renmatix to produce low-cost, large volume quantities of industrial sugar from lignocellulosic biomass (wood, cane bagasse or straw). Renmatix is currently the only commercial player utilizing supercritical hydrolysis as a route to cellulosic sugar production.

Renmatix is working with multiple partners on development of commercial scale facilities with the capability to produce more than 100,000 tons of cellulosic sugars annually. The company has a world-class technical center in Pennsylvania and production operations at the Integrated Plantrose Complex (IPC) in Kennesaw, Georgia and the Feedstock Processing Facility (FPF) in Rome, New York.

In June 2013, Renmatix also entered a joint development agreement (JDA) with UPM, a Finnish pulp, paper and timber manufacturer, to convert woody biomass into low-cost sugar intermediates for subsequent downstream processing into biochemicals.

In December 2013, Renmatix and Virent announced a strategic collaboration to convert affordable cellulosic sugars to renewable chemicals and bio-based packaging materials.

In March 2015, French Energy Group, Total S.A. entered a joint development agreement (JDA) with Renmatix to use the Plantrose technology to extract second-generation sugars from biomass and develop sustainable and profitable biomolecules for products of interest.

Biotechnology Penetration in the Chemical Industry

Year	Value	Penetration
2000 (actual)	$67 billion	5.3%
2005 (actual)	$98 billion	6.7%
2010 (forecast)	$159 billion	9.6%
2025 (projection)	$1000 billion	33%

World Biobased Market Penetration 2010-2025

Chemical Sector	2010	2015
Commodity Chemicals	1-2%	6-10%

Specialty Chemicals	20-25%	45-50%
Fine Chemicals	20-25%	45-50%
Polymers	5-10%	10-20%

In the first quarter of 2013, American Process Inc. announced the start-up of cellu-losic sugar production using their patented AVAP® technology at their demonstra-tion plant in Thomaston, GA. The AVAP process uses ethanol and sulfur dioxide (SO2) to fractionate biomass into it's pure components: cellulose, hemicellulose sugars and lignin. In early 2013 GranBio, a Brazilian pioneer in biofuels and bio-chemicals, completed the acquisition of an equity investment in API. Since that time, API has fermented both five-carbon and six-carbon sugars into high value bio-chemicals and biofuels in partnership with fermentation companies through-out the world.

Applications

Cellulosic sugars are used as renewable resources for biochemical and biofuels in-dustries and can be used to produce intermediates by fermentative processes. The availability of industrial sugars from renewable resources, in sufficient quantities and at a favorable cost enables the products to be cost-competitive to fossil fuel based products.

A 2012 study by Nexant estimates that in the future, it will be possible and potentially economically viable to produce any type of sugar-based chemical product from biomass due to developments in cellulosic processing.

Straw

Large round bales

Bundles of rice straw

Straw is an agricultural by-product, the dry stalks of cereal plants, after the grain and chaff have been removed. Straw makes up about half of the yield of cereal crops such as barley, oats, rice, rye and wheat. It has many uses, including fuel, livestock bedding and fodder, thatching and basket-making. It is usually gathered and stored in a straw bale, which is a bundle of straw tightly bound with twine or wire. Bales may be square, rectangular, or round, depending on the type of baler used.

Pile of "small square" straw bales, sheltered under a tarpaulin

Uses

Current and historic uses of straw include:

- Animal feed

    o Straw may be fed as part of the roughage component of the diet to cattle or horses that are on a near maintenance level of energy requirement. It has a low digestible energy and nutrient content. The heat generated when microorganisms in a herbivore's gut digest straw can be useful in maintaining body temperature in cold climates. Due to the risk of impaction and its poor nutrient profile, it should always be restricted to part of the diet. It may be fed as it is, or chopped into short lengths, known as chaff.

- Basketry

  - Bee skeps and linen baskets are made from coiled and bound together continuous lengths of straw. The technique is known as lip work.

- Bedding: humans or livestock

  - The straw-filled mattress, also known as a palliasse, is still used in many parts of the world.

  - It is commonly used as bedding for ruminants and horses. It may be used as bedding and food for small animals, but this often leads to injuries to mouth, nose and eyes as straw is quite sharp.

- Biofuels

  - The use of straw as a carbon-neutral energy source is increasing rapidly, especially for biobutanol. Straw or hay briquettes are a biofuel substitute to coal.

- Biogas

  - Straw has been tested for use in a biogas plant in Aarhus University, Denmark. Straw has been processed as briquettes and feed into a biogas plant to see if higher gas yields were attainable.

- Biomass

  - The use of straw in large-scale biomass power plants is becoming mainstream in the EU, with several facilities already online. The straw is either used directly in the form of bales, or densified into pellets which allows for the feedstock to be transported over longer distances. Finally, torrefaction of straw with pelletisation is gaining attention, because it increases the energy density of the resource, making it possible to transport it still further. This processing step also makes storage much easier, because torrefied straw pellets are hydrophobic. Torrefied straw in the form of pellets can be directly co-fired with coal or natural gas at very high rates and make use of the processing infrastructures at existing coal and gas plants. Because the torrefied straw pellets have superior structural, chemical and combustion properties to coal, they can replace all coal and turn a coal plant into an entirely biomass-fed power station. First generation pellets are limited to a co-firing rate of 15% in modern IGCC plants.

- Construction material:

  - In many parts of the world, straw is used to bind clay and concrete. A mixture of clay and straw, known as cob, can be used as a building material. There are many recipes for making cob.

○ When baled, straw has moderate insulation characteristics (about R-1.5/inch according to Oak Ridge National Lab and Forest Product Lab testing). It can be used, alone or in a post-and-beam construction, to build straw bale houses. When bales are used to build or insulate buildings, the straw bales are commonly finished with earthen plaster. The plastered walls provide some thermal mass, compressive and ductile structural strength, and acceptable fire resistance as well as thermal resistance (insulation), somewhat in excess of North American building code. Straw is an abundant agricultural waste product, and requires little energy to bale and transport for construction. For these reasons, strawbale construction is gaining popularity as part of passive solar and other renewable energy projects.

○ Composite lumber Wheat straw can be used as a polymer filler combined with polymers to produce composite lumber.

○ Enviroboard can be made from straw.

○ Strawblocks

• Crafts

    ○ Corn dollies

    ○ Straw marquetry

    ○ Straw painting

    ○ Straw plaiting

    ○ Scarecrows

    ○ Japanese Traditional Cat's House

Belarusian Straw Dolls

Easter bunny made of Straw

- Erosion control

    o Straw bales are sometimes used for sediment control at construction sites. However, bales are often ineffective in protecting water quality and are maintenance-intensive. For these reasons the U.S. Environmental Protection Agency (EPA) and various state agencies recommend use of alternative sediment control practices where possible, such as silt fences, fiber rolls and geotextiles.

    o Burned area emergency response

    o Ground cover

    o In-stream check dams

- Hats

    o There are several styles of straw hats that are made of woven straw.

    o Many thousands of women and children in England (primarily in the Luton district of Bedfordshire), and large numbers in the United States (mostly Massachusetts), were employed in plaiting straw for making hats. By the late 19th century, vast quantities of plaits were being imported to England from Canton in China, and in the United States most of the straw plait was imported.

    o A fiber analogous to straw is obtained from the plant Carludovica palmata, and is used to make Panama hats.

- Horticulture

    o Straw is used in cucumber houses and for mushroom growing.

    o In Japan, certain trees are wrapped with straw to protect them from the effects of a hard winter as well as to use them as a trap for parasite insects.

- o It is also used in ponds to reduce algae by changing the nutrient ratios in the water.

- o The soil under strawberries is covered with straw to protect the ripe berries from dirt, and straw is also used to cover the plants during winter to prevent the cold from killing them.

- o Straw also makes an excellent mulch.

- Packaging

    - o Straw is resistant to being crushed and therefore makes a good packing material. A company in France makes a straw mat sealed in thin plastic sheets.

    - o Straw envelopes for wine bottles have become rarer, but are still to be found at some wine merchants.

    - o Wheat straw is also used in compostable food packaging such as compostable plates. Packaging made from wheat straw can be certified compostable and will biodegrade in a commercial composting environment.

- Paper

    - o Straw can be pulped to make paper.

Belarusian Straw Bird

- Rope

    - o Rope made from straw was used by thatchers, in the packaging industry and even in iron foundries.

- Shoes

    o   Koreans wear Jipsin, sandals made of straw.

    o   In some parts of Germany like Black Forest and Hunsrück people wear straw shoes at home or at carnival.

- Targets

    o   Heavy gauge straw rope is coiled and sewn tightly together to make archery targets. This is no longer done entirely by hand, but is partially mechanised. Sometimes a paper or plastic target is set up in front of straw bales, which serve to support the target and provide a safe backdrop.

- Thatching

    o   Thatching uses straw, reed or similar materials to make a waterproof, lightweight roof with good insulation properties. Straw for this purpose (often wheat straw) is grown specially and harvested using a reaper-binder.

Safety

Dried straw presents a fire hazard that can ignite easily if exposed to sparks or an open flame. It can also trigger Allergic rhinitis in people who are hypersensitive to airborne allergens such as straw dust.

Research

In addition to its current and historic uses, straw is being investigated as a source of fine chemicals including alkaloids, flavonoids, lignins, phenols, and steroids.

Woodchips

Woodchips are a medium-sized solid material made by cutting, or chipping, larger pieces of wood.

Woodchips may be used as a biomass solid fuel and are raw material for producing wood pulp. They may also be used as an organic mulch in gardening, landscaping, restoration ecology, bioreactors for denitrification and mushroom cultivation. According to the different chemical and mechanical properties of the masses, the wood logs are mostly peeled, and the bark chips and the wood chips are processed in different processes. The process of making wood chips is called woodchipping and is done with a woodchipper.

Large woodchipper (Europe Chippers model C1175).
This type of machine is used to chip large pieces of wood

Raw Materials

Woodchips waiting to be loaded at Albany Port in Western Australia

The raw materials of wood chips can be pulpwood, wood plantations, waste wood and residuals from construction, agriculture, landscaping, logging, and sawmills and locally grown and harvested fuel crops.

Production

A woodchipper is a machine used for reducing wood to smaller pieces. There are several types of woodchippers depending of the further processing of the woodchips. For industrial use, the woodchippers are large, stationary installations.

Pulp and Paper Industry

Wood chips used for chemical pulp must be relatively uniform in size and free of bark. The optimum size varies with the wood species. It is important to avoid damage to the wood fibres as this is important for the pulp properties. For roundwood it is most common to use disk chippers. A typical size of the disk is 2.0 - 3.5 m in diameter, 10 – 25 cm in thickness and weight is up to 30 tons. The disk is fitted with 4 to 16 knives and driven with motors of ½ - 2 MW. Drum chippers are normally used for wood residuals from saw mills or other wood industry.

Methods of Conveyance

There are four potential methods to move woodchips: pneumatic, conveyor belt, hopper with direct chute, batch system (manual conveyance).

Applications

Woodchips are used primarily as a raw material for technical wood processing. In industry, processing of bark chips is often separated after peeling the logs due to different chemical properties.

Wood Pulp

Only the heartwood and sapwood are useful for making pulp. Bark contains relatively few useful fibres and is removed and used as fuel to provide steam for use in the pulp mill. Most pulping processes require that the wood be chipped and screened to provide uniform sized chips.

Mulch

Woodchipping is also used to produce landscape and garden woodchips mulch. It is used for water conservation, weed control, reducing and preventing soil erosion, and for supporting germination of native seeds and acorns in habitat revegetation-ecological restoration projects. As the ramial chipped wood decomposes it improves the soil structure, permeability, bioactivity, and nutrient availability. Woodchips when used as a mulch are at least three inches thick.

Playground Surfacing

Wood chips can be reprocessed into an extremely effective playground surfacing material, or impact-attenuation surface. When used as playground surfacing (soft fall, cushion fall, or play chip, as it is sometimes known), woodchips can be very effective in lessening the impact of falls from playground equipment. When spread to depths of one foot (30 centimeters) playground wood chips can be effective at reducing impacts in falls up to 11 feet (3 meters). Playground woodchip is also an environmentally friendly alternative to rubber type playground surfaces.

Bioreactor

Woodchip piles at the edge of a field can inhibit nitrates from running off into water tiles. They are a simple measure for farmers to reduce nitrate pollution of the watershed without them having to change their land management practice. A 2011 study showed that most of the nitrate removal was due to heterotrophic denitrification. A 2013 experiment from Ireland showed that after 70 days of startup, a woodchip pile loaded with liquid pig manure at 5 L/m²/day removed an average of 90% of nitrate in the form of ammonium after one month. A January 2015 study from Ohio State University showed very low nitrogen gas, i.e. greenhouse gas emissions from nitrate transformation under the anaerobic conditions of the wood chip bioreactor. Scientists constructed a model for water flow and nitrate removal kinetics which can be used to design denitrification beds. It is unknown if other nutrients like phosphorus or pathogens are affected by the bioreactor as well.

Fuel

Woodchips have been traditionally used as solid fuel for space heating or in energy plants to generate electric power from renewable energy. The main source of forest chips in Europe and in most of the countries have been logging residues. It is expected that the shares of stumps and roundwood will increase in the future. As of 2013 in the EU, the estimates for biomass potential for energy, available under current conditions including sustainable use of the forest as well as providing wood to the traditional forest sectors, are: 277 million m³, for above ground biomass and 585 million m³ for total biomass.

The newer fuel systems for heating use either woodchips or wood pellets. The advantage of woodchips is cost, the advantage of wood pellets is the controlled fuel value. The use of woodchips in automated heating systems, is based on a robust technology.

The size of the woodchips is particularly important when burning woodchip in small plants. Unfortunately there are not many standards to decide the fractions of woodchip. One standard is the GF60 which is commonly used in smaller plants, including small industries, villas, and apartment buildings. "GF60" is known as "Fine, dry, small chips". The requirements for GF60 are that the moisture is between 10–30% and the fractions of the woodchips are distributed as follows: 0–3.5mm: <8%, 3.5–30mm: <7%, 30–60mm: 80–100%, 60–100mm: <3%, 100–120mm: <2%.

The energy content in one cubic metre is normally higher than in one cubic metre wood logs, but can vary greatly depending on moisture. The moisture is decided by the handling of the raw material. If the trees is taken down in the winter and is left to dry under the summer, with tears in the bark and covered so rain can't reach to them and then is chipped in the fall the woodchip will get an moisture of approx. 20–25%. The energy content is then approx 3.5–4.5kWh/kg (~150–250 kg/cubic metre).

Coal power plants have been converted to run on woodchips, which is fairly straightforward to do, since they both use an identical steam turbine heat engine, and the cost of woodchip fuel is comparable to coal.

Solid biomass is an attractive fuel for addressing the concerns of the energy crisis and climate change, since the fuel is affordable, widely available, close to carbon neutral and thus climate-neutral in terms of carbon dioxide (CO_2), since in the ideal case only the carbondioxide which was drawn in during the tree's growth and stored in the wood is released into the atmosphere again. It is sustainable as long as crops are allowed to regrow; In most cases, biomass is not carbon neutral as wood is not regrown and the efficiency of biomass operations produce more pollutants than the processes they replace.

Waste and Emissions

Compared to the solid waste disposal problems of coal and nuclear fuels, woodchip fuel's waste disposal problems are less grave; in a study from 2001 fly ash from wood chip combustion had 28.6 mg cadmium/kg dry matter. Compared to fly ash from burning of straw, cadmium was bound more heavily, with only small amounts of cadmium leached. It was speciated as a form of cadmium oxide, cadmium silicate ($CdSiO_3$); authors noted that adding it to agricultural or forest soils in the long-term could cause a problem with accumulation of cadmium.

Like coal, wood combustion is a known source of mercury emissions, particularly in northern climates during winter. The mercury is both gaseous as elemental mercury (especially when wood pellets are burned) or mercury oxide, and solid PM2.5 particulate matter when untreated wood is used.

When wood burning is used for space heating, indoor emissions of 1,3-butadiene, benzene, formaldehyde and acetaldehyde, which are suspected or known carcinogenic compounds, are elevated. The cancer risk from these after exposure to wood smoke is estimated to be low in developed countries .

Certain techniques for burning woodchips result in the production of biochar - effectively charcoal - which can be either utilised as charcoal, or returned to the soil, since wood ash can be used as a mineral-rich plant fertilizer. The latter method can result in an effectively carbon-negative system, as well as acting as a very effective soil conditioner, enhancing water and nutrient retention in poor soils.

Automated Handling of Solid Fuel

Unlike the smooth, uniform shape of manufactured wood pellets, woodchip sizes vary and are often mixed with twigs and sawdust. This mixture has a higher probability of jamming in small feed mechanisms. Thus, sooner or later, one or more jams is likely to occur. This reduces the reliability of the system, as well as increasing maintenance costs. Despite what some pellet stove manufacturers may say, researchers who are experienced with woodchips, say they are not compatible with the 2 inch (5 cm) auger used in pellet stoves.

Micro Combined Heat and Power

Wood is occasionally used to power engines, such as steam engines, Stirling engines, and Otto engines running on woodgas. As of 2008, these systems are rare, but as technology and the need for it develops, it is likely to be more common in the future. For the time being, wood can be increasingly used for heating applications. This will reduce the demand for heating oil, and thereby allow a greater percentage of fuel oil to be used for applications such as internal combustion engines, which are less compatible with wood based fuel and other solid biomass fuels. Heating applications generally do not require refined or processed fuels, which are almost always more expensive.

Comparison to Other Fuels

Woodchips are similar to wood pellets, in that the movement and handling is more amenable to automation than cord wood, particularly for smaller systems. Woodchips are less expensive than wood pellets and are theoretically more energy efficient than pellets, because less energy is required for manufacturing, processing, and transportation; however, this assumes that they are consumed in an appropriately designed burner, and as of 2008, these are mostly only available in large systems designed for commercial or institutional use, which have been very successful in terms of performance, cost, reliability, and efficiency.

Woodchips are less expensive than cord wood, because the harvesting is faster and more highly automated. Woodchips are of greater supply, partly because all parts of a tree can be chipped, whereas small limbs and branches can require much labor convert to cord wood. Cordwood generally needs to be "seasoned" or "dry" before it can be burned cleanly and efficiently. On the other hand, woodchip systems are typically designed to cleanly and efficiently burn "green chips" with very high moisture content of 43–47% (wet basis).

Environmental Aspects

If woodchips are harvested through sustainable forestry practices, then this is considered a source of renewable energy. On the other hand, harvesting practices, such as clearcutting large areas, are highly damaging to forest ecosystems.

Theoretically, whole-tree chip harvesting does not have as high a solar energy efficiency compared to short rotation coppice; however, it can be an energy-efficient and low-cost method of harvesting. In some cases, this practice may be controversial when whole-tree harvesting may often be associated with clear cutting and perhaps other questionable forestry practices.

Waste Processing

Woodchips, and bark chips, can be used as bulking agents in industrial composting of municipal biodegradeable waste, particularly biosolids.

Woodchip biomass does not have the waste disposal issues of coal and nuclear power, since wood ash can be used directly as a mineral-rich plant fertilizer.

Forest Fire Prevention

Woodchip harvesting can be used in concert with creating man made firebreaks, which are used as barriers to the spread of wildfire. Undergrowth coppice is ideal for chipping, and larger trees may be left in place to shade the forest floor and reduce the rate of fuel accumulation.

Market Products, Supply and Demand

Currently, domestic or residential sized systems are not available in products for sale on the general market. Homemade devices have been produced, that are small-scale, clean-burning, and efficient for woodchip fuels. Much of the research activity to date, has consisted of small budget projects that are self-funded. The majority of funding for energy research has been for liquid biofuels.

United States

"Wood chip costs usually depend on such factors as the distance from the point of delivery, the type of material (such as bark, sawmill residue or whole-tree chips), demand by other markets and how the wood fuel is transported. Chips delivered directly to the (powerplant) station by truck are less expensive than those delivered ... and shipped by railcar. The range of prices is typically between US$18 to US$30 per (wet)-ton delivered."

In 2006, prices were US$15 and US$30 per wet-ton in the northeast.

In the 20 years leading up to 2008, prices have fluctuated between US$60–70/oven-dry metric ton (odmt) in the southern states, and between US$60/odmt and US$160/odmt in the Northwest.

Europe

Large woodchipper in Germany

In several well wooded European countries (e.g. Austria, Finland, Germany, Sweden) wood chips are becoming an alternative fuel for family homes and larger buildings due to the abundant availability of wood chips, which result in low fuel costs. The European Union is promoting wood chips for energy production in the EU Forest action plan 2007-2011. The total long term potential of wood chips in the EU it is estimated to be 913 million m3.

Japan

After a long period of negative scores, the demand of wood chip for paper manufacturing started increasing again. Starting in the last quarter of 2013, orders for printing paper and card board increased before the consumption tax increase then by weakening yen, import of papers like copy paper decreases and export of paper increases, which stimulate paper production in Japan. Softwood chip prices from the United States increased by 12% compared to October 2013 and softwood chip prices from Australia increased by 7%.

Bioconversion of Biomass to Mixed Alcohol Fuels

The bioconversion of biomass to mixed alcohol fuels can be accomplished using the MixAlco process. Through bioconversion of biomass to a mixed alcohol fuel, more energy from the biomass will end up as liquid fuels than in converting biomass to ethanol by yeast fermentation.

The process involves a biological/chemical method for converting any biodegradable material (e.g., urban wastes, such as municipal solid waste, biodegradable waste, and sewage sludge, agricultural residues such as corn stover, sugarcane bagasse, cotton gin trash, manure) into useful chemicals, such as carboxylic acids (e.g., acetic, propionic, butyric acid), ketones (e.g., acetone, methyl ethyl ketone, diethyl ketone) and biofuels, such as a mixture of primary alcohols (e.g., ethanol, propanol, n-butanol) and/or a mixture of secondary alcohols (e.g., isopropanol, 2-butanol, 3-pentanol). Because of the many products that can be economically produced, this process is a true biorefinery.

Pilot Plant (College Station, Texas)

The process uses a mixed culture of naturally occurring microorganisms found in natural habitats such as the rumen of cattle, termite guts, and marine and terrestrial swamps to anaerobically digest biomass into a mixture of carboxylic acids produced during the acidogenic and acetogenic stages of anaerobic digestion, however with the inhibition of the methanogenic final stage. The more popular methods for production of ethanol and cellulosic ethanol use enzymes that must be isolated first to be added to the biomass and thus convert the starch or cellulose into simple sugars, followed then by yeast fermentation into ethanol. This process does not need the addition of such enzymes as these microorganisms make their own.

As the microoganisms anaerobically digest the biomass and convert it into a mixture of carboxylic acids, the pH must be controlled. This is done by the addition of a buffering agent (e.g., ammonium bicarbonate, calcium carbonate), thus yielding a mixture of carboxylate salts. Methanogenesis, being the natural final stage of anaerobic digestion, is inhibited by the presence of the ammonium ions or by the addition of an inhibitor (e.g., iodoform). The resulting fermentation broth contains the produced carboxylate salts that must be dewatered. This is achieved efficiently by vapor-compression evaporation. Further chemical refining of the dewatered fermentation broth may then take place depending on the final chemical or biofuel product desired.

The condensed distilled water from the vapor-compression evaporation system is recycled back to the fermentation. On the other hand, if raw sewage or other waste water with high BOD in need of treatment is used as the water for the fermentation, the condensed distilled water from the evaporation can be recycled back to the city or to the original source of the high-BOD waste water. Thus, this process can also serve as a water treatment facility, while producing valuable chemicals or biofuels.

Because the system uses a mixed culture of microorganisms, besides not needing any enzyme addition, the fermentation requires no sterility or aseptic conditions, making this front step in the process more economical than in more popular methods for the production of cellulosic ethanol. These savings in the front end of the process, where volumes are large, allows flexibility for further chemical transformations after dewatering, where volumes are small.

Carboxylic Acids

Carboxylic acids can be regenerated from the carboxylate salts using a process known as "acid springing". This process makes use of a high-molecular-weight tertiary amine (e.g., trioctylamine), which is switched with the cation (e.g., ammonium or calcium). The resulting amine carboxylate can then be thermally decomposed into the amine itself, which is recycled, and the corresponding carboxylic acid. In this way, theoretically, no chemicals are consumed or wastes produced during this step.

Ketones

There are two methods for making ketones. The first one consists on thermally converting calcium carboxylate salts into the corresponding ketones. This was a common method for making acetone from calcium acetate during World War I. The other method for making ketones consists on converting the vaporized carboxylic acids on a catalytic bed of zirconium oxide.

Alcohols

Primary Alcohols

The undigested residue from the fermentation may be used in gasification to make hydrogen (H_2). This H_2 can then be used to hydrogenolyze the esters over a catalyst (e.g., copper chromite), which are produced by esterifying either the ammonium carboxylate salts (e.g., ammonium acetate, propionate, butyrate) or the carboxylic acids (e.g., acetic, propionic, butyric acid) with a high-molecular-weight alcohol (e.g., hexanol, heptanol). From the hydrogenolysis, the final products are the high-molecular-weight alcohol, which is recycled back to the esterification, and the corresponding primary alcohols (e.g., ethanol, propanol, butanol).

Secondary Alcohols

The secondary alcohols (e.g., isopropanol, 2-butanol, 3-pentanol) are obtained by hydrogenating over a catalyst (e.g., Raney nickel) the corresponding ketones (e.g., acetone, methyl ethyl ketone, diethyl ketone).

Drop-in Biofuels

The primary or secondary alcohols obtained as described above may undergo conversion to drop-in biofuels, fuels which are compatible with current fossil fuel infrastructure such as biogasoline, green diesel and bio-jet fuel. Such is done by subjecting the alcohols to dehydration followed by oligomerization using zeolite catalysts in a manner similar to the methanex process, which used to produce gasoline from methanol in New Zealand.

Acetic Acid Versus Ethanol

Cellulosic-ethanol manufacturing plants are bound to be net exporters of electricity because a large portion of the lignocellulosic biomass, namely lignin, remains undigested and it must be burned, thus producing electricity for the plant and excess electricity for the grid. As the market grows and this technology becomes more widespread, coupling the liquid fuel and the electricity markets will become more and more difficult.

Acetic acid, unlike ethanol, is biologically produced from simple sugars without the production of carbon dioxide:

$$C_6H_{12}O_6 \quad \rightarrow \quad 2\,CH_3CH_2OH \quad + \quad 2\,CO_2$$

(Biological production of ethanol)

$$C_6H_{12}O_6 \quad \rightarrow \quad 3\,CH_3COOH$$

(Biological production of acetic acid)

Because of this, on a mass basis, the yields will be higher than in ethanol fermentation. If then, the undigested residue (mostly lignin) is used to produce hydrogen by gasification, it is ensured that more energy from the biomass will end up as liquid fuels rather than excess heat/electricity.

$$3\,CH_3COOH \quad + \quad 6\,H_2 \quad \rightarrow \quad 3\,CH_3CH_2OH \quad + \quad 3\,H_2O$$

(Hydrogenation of acetic acid)

$$C_6H_{12}O_6 \text{ (from cellulose)} \quad + \quad 6\,H_2 \text{ (from lignin)} \quad \rightarrow \quad 3\,CH_3CH_2OH \quad + \quad 3\,H_2O$$

(Overall reaction)

A more comprehensive description of the economics of each of the fuels is given on the pages alcohol fuel and ethanol fuel, more information about the economics of various systems can be found on the central page biofuel.

Stage of Development

The system has been in development since 1991, moving from the laboratory scale (10 g/day) to the pilot scale (200 lb/day) in 2001. A small demonstration-scale plant (5 ton/day) has been constructed and is under operation and a 220 ton/day demonstration plant is expected in 2012.

Cogeneration

Cogeneration or combined heat and power (CHP) is the use of a heat engine or power station to generate electricity and useful heat at the same time. Trigeneration or combined cooling, heat and power (CCHP) refers to the simultaneous generation of electricity and useful heating and cooling from the combustion of a fuel or a solar heat collector.

Cogeneration is a thermodynamically efficient use of fuel. In separate production of electricity, some energy must be discarded as waste heat, but in cogeneration some of this thermal energy is put to use. All thermal power plants emit heat during electricity generation, which can be released into the natural environment through cooling towers, flue gas, or by other means. In contrast, CHP captures some or all of the by-product for heating, either very close to the plant, or—especially in Scandinavia and Eastern Europe—as hot water for district heating with temperatures ranging from approximately 80 to 130 °C. This is also called combined heat and power district heating (CHPDH). Small CHP plants are an example of decentralized energy. By-product heat at moderate temperatures (100–180 °C, 212–356 °F) can also be used in absorption refrigerators for cooling.

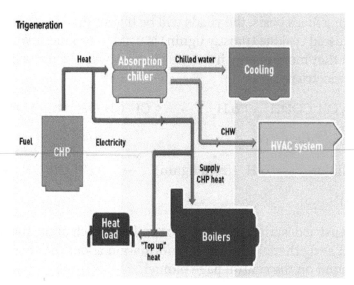

Trigeneration cycle

The supply of high-temperature heat first drives a gas or steam turbine-powered generator and the resulting low-temperature waste heat is then used for water or space heating as described in cogeneration. At smaller scales (typically below 1 MW) a gas engine or diesel engine may be used. Trigeneration differs from cogeneration in that the waste heat is used for both heating and cooling, typically in an absorption refrigerator. CCHP systems can attain higher overall efficiencies than cogeneration or traditional power plants. In the United States, the application of trigeneration in buildings is called building cooling, heating and power (BCHP). Heating and cooling output may operate concurrently or alternately depending on need and system construction.

Cogeneration was practiced in some of the earliest installations of electrical generation. Before central stations distributed power, industries generating their own power used exhaust steam for process heating. Large office and apartment buildings, hotels and stores commonly generated their own power and used waste steam for building heat. Due to the high cost of early purchased power, these CHP operations continued for many years after utility electricity became available.

Overview

Thermal power plants (including those that use fissile elements or burn coal, petroleum, or natural gas), and heat engines in general, do not convert all of their thermal energy into electricity. In most heat engines, slightly more than half is lost as excess heat. By capturing the ex-cess heat, CHP uses heat that would be wasted in a conventional power plant, potential-ly reaching an efficiency of up to 80%, for the best conventional plants. This means that less fuel needs to be consumed to produce the same amount of useful energy.

Masnedø CHP power station in Denmark. This station burns straw as fuel. The adjacent greenhouses are heated by district heating from the plant.

Steam turbines for cogeneration are designed for *extraction* of steam at lower pressures after it has passed through a number of turbine stages, or they may be designed for final exhaust at *back pressure* (non-condensing), or both. A typical power generation turbine in a paper mill may have extraction pressures of 160 psig (1.103 MPa) and 60 psig (0.41 MPa). A typical back pressure may be 60 psig (0.41 MPa). In practice these pressures are custom designed for each facility. The extracted or exhaust steam is used for process heating, such as drying paper, evaporation, heat for chemical reactions or distillation. Steam at ordinary process heating conditions still has a considerable amount of enthalpy that could be used for power generation, so cogeneration has lost opportunity cost. Conversely, simply generating steam at process pressure instead of high enough pressure to generate power at the top end also has lost opportunity cost. The capital and operating cost of high pressure boilers, turbines and generators are substantial, and this equipment is normally operated continuously, which usually limits self-generated power to large-scale operations.

A cogeneration plant in Metz, France. The 45MW boiler uses waste wood biomass as energy source, and provides electricity and heat for 30,000 dwellings.

Some tri-cycle plants have used a combined cycle in which several thermodynamic cycles produced electricity, then a heating system was used as a condenser of the power plant's bottoming cycle. For example, the RU-25 MHD generator in Moscow heated a boiler for a conventional steam powerplant, whose condensate was then used for space heat. A more modern system might use a gas turbine powered by natural gas, whose exhaust powers a steam plant, whose condensate provides heat. Tri-cycle plants can have thermal efficiencies above 80%.

The viability of CHP (sometimes termed utilisation factor), especially in smaller CHP installations, depends on a good baseload of operation, both in terms of an on-site (or near site) electrical demand and heat demand. In practice, an exact match between the heat and electricity needs rarely exists. A CHP plant can either meet the need for heat (*heat driven operation*) or be run as a power plant with some use of its waste heat, the latter being less advantageous in terms of its utilisation factor and thus its overall efficiency. The viability can be greatly increased where opportunities for Trigeneration exist. In such cases, the heat from the CHP plant is also used as a primary energy source to deliver cooling by means of an absorption chiller.

CHP is most efficient when heat can be used on-site or very close to it. Overall efficiency is reduced when the heat must be transported over longer distances. This requires heavily insulated pipes, which are expensive and inefficient; whereas electricity can be transmitted along a comparatively simple wire, and over much longer distances for the same energy loss.

A car engine becomes a CHP plant in winter when the reject heat is useful for warming the interior of the vehicle. The example illustrates the point that deployment of CHP depends on heat uses in the vicinity of the heat engine.

Thermally enhanced oil recovery (TEOR) plants often produce a substantial amount of excess electricity. After generating electricity, these plants pump leftover steam into heavy oil wells so that the oil will flow more easily, increasing production. TEOR cogeneration plants in Kern County, California produce so much electricity that it cannot all be used locally and is transmitted to Los Angeles.

CHP is one of the most cost-efficient methods of reducing carbon emissions from heating systems in cold climates and is recognized to be the most energy efficient method of transforming energy from fossil fuels or biomass into electric power. Cogeneration plants are commonly found in district heating systems of cities, central heating systems from buildings, hospitals, prisons and are commonly used in the industry in thermal production processes for process water, cooling, steam production or CO_2 fertilization.

Types of Plants

Topping cycle plants primarily produce electricity from a steam turbine. The exhausted steam is then condensed and the low temperature heat released from this condensation is utilized for e.g. district heating or water desalination.

Bottoming cycle plants produce high temperature heat for industrial processes, then a waste heat recovery boiler feeds an electrical plant. Bottoming cycle plants are only used when the industrial process requires very high temperatures such as furnaces for glass and metal manufacturing, so they are less common.

Large cogeneration systems provide heating water and power for an industrial site or an entire town. Common CHP plant types are:

- Gas turbine CHP plants using the waste heat in the flue gas of gas turbines. The fuel used is typically natural gas.

- Gas engine CHP plants use a reciprocating gas engine which is generally more competitive than a gas turbine up to about 5 MW. The gaseous fuel used is normally natural gas. These plants are generally manufactured as fully packaged units that can be installed within a plantroom or external plant compound with simple connections to the site's gas supply, electrical distribution network and heating systems.

- Biofuel engine CHP plants use an adapted reciprocating gas engine or diesel engine, depending upon which biofuel is being used, and are otherwise very similar in design to a Gas engine CHP plant. The advantage of using a biofuel is one of reduced hydrocarbon fuel consumption and thus reduced carbon emissions. These plants are generally manufactured as fully packaged units that can be installed within a plantroom or external plant compound with simple connections to the site's electrical distribution and heating systems. Another variant is the wood gasifier CHP plant whereby a wood pellet or wood chip biofuel is gasified in a zero oxygen high temperature environment; the resulting gas is then used to power the gas engine.

- Combined cycle power plants adapted for CHP

- Molten-carbonate fuel cells and solid oxide fuel cells have a hot exhaust, very suitable for heating.

- Steam turbine CHP plants that use the heating system as the steam condenser for the steam turbine.

- Nuclear power plants, similar to other steam turbine power plants, can be fitted with extractions in the turbines to bleed partially expanded steam to a heating system. With a heating system temperature of 95 °C it is possible to extract about 10 MW heat for every MW electricity lost. With a temperature of 130 °C the gain is slightly smaller, about 7 MW for every MWe lost.

Smaller cogeneration units may use a reciprocating engine or Stirling engine. The heat is removed from the exhaust and radiator. The systems are popular in small sizes because small gas and diesel engines are less expensive than small gas- or oil-fired steam-electric plants.

Some cogeneration plants are fired by biomass, or industrial and municipal solid waste. Some CHP plants utilize waste gas as the fuel for electricity and heat generation. Waste gases can be gas from animal waste, landfill gas, gas from coal mines, sewage gas, and combustible industrial waste gas.

Some cogeneration plants combine gas and solar photovoltaic generation to further improve technical and environmental performance. Such hybrid systems can be scaled down to the building level and even individual homes. More recent results show that solar photovoltaic + battery + CHP hybrid systems are technically viable in the continental U.S. to reduce consumer costs, while reducing energy- and electricity-related greenhouse gas emissions.

MicroCHP

Micro combined heat and power or 'Micro cogeneration" is a so-called distributed energy resource (DER). The installation is usually less than 5 kW_e in a house or small business. Instead of burning fuel to merely heat space or water, some of the energy is converted to electricity in addition to heat. This electricity can be used within the home or business or, if permitted by the grid management, sold back into the electric power grid.

Delta-ee consultants stated in 2013 that with 64% of global sales the fuel cell micro-combined heat and power passed the conventional systems in sales in 2012. 20.000 units were sold in Japan in 2012 overall within the Ene Farm project. With a Lifetime of around 60,000 hours. For PEM fuel cell units, which shut down at night, this equates to an estimated lifetime of between ten and fifteen years. For a price of $22,600 before installation. For 2013 a state subsidy for 50,000 units is in place.

The development of small-scale CHP systems has provided the opportunity for in-house power backup of residential-scale photovoltaic (PV) arrays. The results of a 2011 study show that a PV+CHP hybrid system not only has the potential to radically reduce energy waste in the status quo electrical and heating systems, but it also enables the share of solar PV to be expanded by about a factor of five. In some regions, in order to reduce waste from excess heat, an absorption chiller has been proposed to utilize the CHP-produced thermal energy for cooling of PV-CHP system. These trigeneration+-photovoltaic systems have the potential to save even more energy and further reduce emissions compared to conventional sources of power, heating and cooling.

MicroCHP installations use five different technologies: microturbines, internal combustion engines, stirling engines, closed cycle steam engines and fuel cells. One author indicated in 2008 that MicroCHP based on Stirling engines is the most cost effective of the so-called microgeneration technologies in abating carbon emissions; A 2013 UK report from Ecuity Consulting stated that MCHP is the most cost-effective method of utilising gas to generate energy at the domestic level. however, advances in reciprocation engine technology are adding efficiency to CHP plant, particularly in the biogas field. As both MiniCHP and CHP have been shown to reduce emissions they could play a large role in the field of CO_2 reduction from buildings, where more than 14% of emissions can be saved using CHP in buildings. The ability to reduce emissions is particularly strong for new communities in emission intensive grids that utilize a combination of CHP and photovoltaic systems.

Trigeneration

A plant producing electricity, heat and cold is called a trigeneration or polygeneration plant. Cogeneration systems linked to absorption chillers use waste heat for refrigeration.

Combined Heat and Power District Heating

In the United States, Consolidated Edison distributes 66 billion kilograms of 350 °F (180 °C) steam each year through its seven cogeneration plants to 100,000 buildings in Manhattan—the biggest steam district in the United States. The peak delivery is 10 million pounds per hour (or approximately 2.5 GW).

Industrial CHP

Cogeneration is still common in pulp and paper mills, refineries and chemical plants. In this "industrial cogeneration/CHP", the heat is typically recovered at higher temperatures (above 100 deg C) and used for process steam or drying duties. This is more valuable and flexible than low-grade waste heat, but there is a slight loss of power generation. The increased focus on sustainability has made industrial CHP more attractive, as it substantially reduces carbon footprint compared to generating steam or burning fuel on-site and importing electric power from the grid.

Utility Pressures Versus Self Generating Industrial

Industrial cogeneration plants normally operate at much lower boiler pressures than utilities. Among the reasons are: 1) Cogeneration plants face possible contamination of returned condensate. Because boiler feed water from cogeneration plants has much lower return rates than 100% condensing power plants, industries usually have to treat proportionately more boiler make up water. Boiler feed water must be completely oxygen free and de-mineralized, and the higher the pressure the more critical the level of purity of the feed water. 2) Utilities are typically larger scale power than industry, which helps offset the higher capital costs of high pressure. 3) Utilities are less likely to have sharp load swings than industrial operations, which deal with shutting down or starting up units that may represent a significant percent of either steam or power demand.

Heat Recovery Steam Generators

A heat recovery steam generator (HRSG) is a steam boiler that uses hot exhaust gases from the gas turbines or reciprocating engines in a CHP plant to heat up water and generate steam. The steam, in turn, drives a steam turbine or is used in industrial processes that require heat.

HRSGs used in the CHP industry are distinguished from conventional steam generators by the following main features:

- The HRSG is designed based upon the specific features of the gas turbine or reciprocating engine that it will be coupled to.

- Since the exhaust gas temperature is relatively low, heat transmission is accomplished mainly through convection.

- The exhaust gas velocity is limited by the need to keep head losses down. Thus, the transmission coefficient is low, which calls for a large heating surface area.

- Since the temperature difference between the hot gases and the fluid to be heated (steam or water) is low, and with the heat transmission coefficient being low as well, the evaporator and economizer are designed with plate fin heat exchangers.

Comparison with a Heat Pump

A heat pump may be compared with a CHP unit, in that for a condensing steam plant, as it switches to produce heat, then electrical generation becomes unavailable, just as the power used in a heat pump becomes unavailable. Typically for every unit of electrical power lost, then about 6 units of heat are made available at about 90 °C. Thus CHP has an effective Coefficient of Performance (COP) compared to a heat pump of 6. It is noteworthy that the unit for the CHP is lost at the high voltage network and therefore incurs no losses, whereas the heat pump unit is lost at the low voltage part of the network and incurs on average a 6% loss. Because the losses are proportional to the square of the current, during peak periods losses are much higher than this and it is likely that widespread (i.e. city-wide application of heat pumps) would cause overloading of the distribution and transmission grids unless they are substantially reinforced.

It is also possible to run a heat driven operation combined with a heat pump, where the excess electricity (as heat demand is the defining factor on utilization) is used to drive a heat pump. As heat demand increases, more electricity is generated to drive the heat pump, with the waste heat also heating the heating fluid.

Distributed Generation

Trigeneration has its greatest benefits when scaled to fit buildings or complexes of buildings where electricity, heating and cooling are perpetually needed. Such installations include but are not limited to: data centers, manufacturing facilities, universities, hospitals, military complexes, and schools. Localized trigeneration has addition benefits as described by distributed generation. Redundancy of power in mission critical applications, lower power usage costs and the ability to sell electrical power back to the local utility are a few of the major benefits. Even for small buildings such as individual family homes trigeneration systems provide benefits over cogeneration because of increased energy utilization. This increased efficiency can also provide significant reduced greenhouse gas emissions, particularly for new communities.

Most industrial countries generate the majority of their electrical power needs in large centralized facilities with capacity for large electrical power output. These plants have excellent economies of scale, but usually transmit electricity long distances resulting in sizable losses, negatively affect the environment. Large power plants can use cogeneration or trigeneration systems only when sufficient need exists in immediate geographic vicinity for an industrial complex, additional power plant or a city. An example of cogeneration with trigeneration applications in a major city is the New York City steam system.

Thermal Efficiency

Every heat engine is subject to the theoretical efficiency limits of the Carnot cycle. When the fuel is natural gas, a gas turbine following the Brayton cycle is typically used. Mechanical energy from the turbine drives an electric generator. The low-grade (i.e. low temperature) waste heat rejected by the turbine is then applied to space heating or cooling or to industrial processes. Cooling is achieved by passing the waste heat to an absorption chiller.

Thermal efficiency in a trigeneration system is defined as:

$$\eta_{th} \equiv \frac{W_{out}}{Q_{in}}$$

Where:

η_{th} = Thermal efficiency

W_{out} = Total work output by all systems

Q_{in} = Total heat input into the system

Typical trigeneration models have losses as in any system. The energy distribution below is represented as a percent of total input energy:

Electricity = 45%

Heat + Cooling = 40%

Heat Losses = 13%

Electrical Line Losses = 2%

Conventional central coal- or nuclear-powered power stations convert only about 33% of their input heat to electricity. The remaining 67% emerges from the turbines as low-grade waste heat with no significant local uses so it is usually rejected to the environment. These low conversion efficiencies strongly suggest that productive uses could be found for this waste heat, and in some countries these plants do collect byproduct heat that can be sold to customers.

But if no practical uses can be found for the waste heat from a central power station, e.g., due to distance from potential customers, then moving generation to where the waste heat can find uses may be of great benefit. Even though the efficiency of a small distributed electrical generator may be lower than a large central power plant, the use of its waste heat for local heating and cooling can result in an overall use of the primary fuel supply as great as 80%. This provides substantial financial and environmental benefits.

Costs

Typically, for a gas-fired plant the fully installed cost per kW electrical is around £400/kW ($577 USD), which is comparable with large central power stations.

History

Cogeneration in Europe

A cogeneration thermal power plant in Ferrera Erbognone (PV), Italy

The EU has actively incorporated cogeneration into its energy policy via the CHP Directive. In September 2008 at a hearing of the European Parliament's Urban Lodgment Intergroup, Energy Commissioner Andris Piebalgs is quoted as saying, "security of supply really starts with energy efficiency." Energy efficiency and cogeneration are recognized in the opening paragraphs of the European Union's Cogeneration Directive 2004/08/EC. This directive intends to support cogeneration and establish a method for calculating cogeneration abilities per country. The development of cogeneration has been very uneven over the years and has been dominated throughout the last decades by national circumstances.

The European Union generates 11% of its electricity using cogeneration. However, there is large difference between Member States with variations of the energy savings between 2% and 60%. Europe has the three countries with the world's most intensive cogeneration economies: Denmark, the Netherlands and Finland. Of the 28.46 TWh of electrical power generated by conventional thermal power plants in Finland in 2012, 81.80% was cogeneration.

Other European countries are also making great efforts to increase efficiency. Germany reported that at present, over 50% of the country's total electricity demand could be provided through cogeneration. So far, Germany has set the target to double its electricity cogeneration from 12.5% of the country's electricity to 25% of the country's electricity by 2020 and has passed supporting legislation accordingly. The UK is also actively supporting combined heat and power. In light of UK's goal to achieve a 60% reduction in carbon dioxide emissions by 2050, the government has set the target to source at least 15% of its government electricity use from CHP by 2010. Other UK measures to encourage CHP growth are financial incentives, grant support, a greater regulatory framework, and government leadership and partnership.

According to the IEA 2008 modeling of cogeneration expansion for the G8 countries, the expansion of cogeneration in France, Germany, Italy and the UK alone would effectively double the existing primary fuel savings by 2030. This would increase Europe's savings from today's 155.69 Twh to 465 Twh in 2030. It would also result in a 16% to 29% increase in each country's total cogenerated electricity by 2030.

Governments are being assisted in their CHP endeavors by organizations like COGEN Europe who serve as an information hub for the most recent updates within Europe's energy policy. COGEN is Europe's umbrella organization representing the interests of the cogeneration industry.

The European public–private partnership Fuel Cells and Hydrogen Joint Undertaking Seventh Framework Programme project ene.field deploys in 2017 up 1,000 residential fuel cell Combined Heat and Power (micro-CHP) installations in 12 states. Per 2012 the first 2 installations have taken place.

Cogeneration in the United Kingdom

In the United Kingdom, the Combined Heat and Power Quality Assurance (CHPQA) scheme regulates the combined production of heat and power. CHPQA was introduced in 1996. It defines, through calculation of inputs and outputs, "Good Quality CHP" in terms of the achievement of primary energy savings against conventional separate generation of heat and electricity. Compliance with CHPQA is required for cogeneration installations to be eligible for government subsidies and tax incentives.

Cogeneration in the United States

Perhaps the first modern use of energy recycling was done by Thomas Edison. His 1882 Pearl Street Station, the world's first commercial power plant, was a combined heat and power plant, producing both electricity and thermal energy while using waste heat to warm neighboring buildings. Recycling allowed Edison's plant to achieve approximately 50 percent efficiency.

The 250 MW Kendall Cogeneration Station plant in Cambridge, Massachusetts

By the early 1900s, regulations emerged to promote rural electrification through the construction of centralized plants managed by regional utilities. These regulations not only promoted electrification throughout the countryside, but they also discouraged decentralized power generation, such as cogeneration.

By 1978, Congress recognized that efficiency at central power plants had stagnated and sought to encourage improved efficiency with the Public Utility Regulatory Policies Act (PURPA), which encouraged utilities to buy power from other energy producers.

Diffusion

Cogeneration plants proliferated, soon producing about 8% of all energy in the United States. However, the bill left implementation and enforcement up to individual states, resulting in little or nothing being done in many parts of the country.

The United States Department of Energy has an aggressive goal of having CHP constitute 20% of generation capacity by the year 2030. Eight Clean Energy Application Centers have been established across the nation whose mission is to develop the required technology application knowledge and educational infrastructure necessary to lead "clean energy" (combined heat and power, waste heat recovery and district energy) technologies as viable energy options and reduce any perceived risks associated with their implementation. The focus of the Application Centers is to provide an outreach and technology deployment program for end users, policy makers, utilities, and industry stakeholders.

High electric rates in New England and the Middle Atlantic make these areas of the United States the most beneficial for cogeneration.

Outside of the United States, energy recycling is more common. Denmark is probably the most active energy recycler, obtaining about 55% of its energy from cogeneration and waste heat recovery. Other large countries, including Germany, Russia, and India, also obtain a much higher share of their energy from decentralized sources.

Biorefinery

A biorefinery is a facility that integrates biomass conversion processes and equipment to produce fuels, power, heat, and value-added chemicals from biomass. The biorefinery concept is analogous to today's petroleum refinery, which produce multiple fuels and products from petroleum.

The International Energy Agency Bioenergy Task 42 on Biorefineries has defined biorefining as the sustainable processing of biomass into a spectrum of bio-based products (food, feed, chemicals, materials) and bioenergy (biofuels, power and/or heat).

By producing multiple products, a biorefinery takes advantage of the various components in biomass and their intermediates therefore maximizing the value derived from the biomass feedstock. A biorefinery could, for example, produce one or several low-volume, but high-value, chemical or nutraceutical products and a low-value, but high-volume liquid transportation fuel such as biodiesel or bioethanol. At the same time generating electricity and process heat, through combined heat and power (CHP) technology, for its own use and perhaps enough for sale of electricity to the local utility. The high-value products increase profitability, the high-volume fuel helps meet energy needs, and the power production helps to lower energy costs and reduce greenhouse gas emissions from traditional power plant facilities. Although some facilities exist that can be called bio-refineries, the bio-refinery has yet to be fully realized. Future biorefineries may play a major role in producing chemicals and materials that are traditionally produced from petroleum.

Examples

The fully operational Blue Marble Energy company has multiple biorefineries located in Odessa, WA and Missoula, MT.

Canada's first Integrated Biorefinery, developed on anaerobic digestion technology by Himark BioGas is located in Hairy Hill, Alberta. The Biorefinery utilizes Source Separated Organics from the metro Edmonton region, Open Pen Feedlot Manure, and Food Processing Waste.

Several potential biorefinery examples have been proposed, starting from feedstocks such as tobacco, flax straw and the residues from the production of bioethanol.

Chemrec's technology for black liquor gasification and production of second-generation biofuels such as biomethanol or BioDME is integrated with a host pulp mill and utilizes a major sulfate or sulfite process waste product as feedstock.

Bioasphalt

Bioasphalt is an asphalt alternative made from non-petroleum based renewable resources.

These sources include sugar, molasses and rice, corn and potato starches, natural tree and gum resins, natural latex rubber and vegetable oils, lignin, cellulose, palm oil waste, coconut waste, peanut oil waste, canola oil waste, dried sewerage effluent and so on. Bitumen can also be made from waste vacuum tower bottoms produced in the process of cleaning used motor oils, which are normally burned or dumped into land fills.

Non-petroleum based bitumen binders can be colored, which can reduce the temperatures of road surfaces and reduce the Urban heat islands.

Petroleum, Environmental, and Heat Concerns

Because of concerns over Peak oil, pollution and climate change, as well the oil price increases since 2003, non-petroleum alternatives have become more popular. This has led to the introduction of biobitumen alternatives that are more environmentally friendly and nontoxic.

For millions of people living in and around cities, heat islands are of growing concern. This phenomenon describes urban and suburban temperatures that are 1 to 6 °C (2 to 10 °F) hotter than nearby rural areas. Elevated temperatures can impact communities by increasing peak energy demand, air conditioning costs, air pollution levels, and heat-related illness and mortality. Fortunately, there are common-sense measures that communities can take to reduce the negative effects of heat islands, such as replacing conventional black asphalt road surfaces with the new pigmentable bitumen that gives lighter colors.

History and Implementation

Asphalt made with vegetable oil based binders was patented by Colas SA in France in 2004.

A number of homeowners seeking an environmentally friendly alternative to asphalt for paving have experimented with waste vegetable oil as a binder for driveways and parking areas in single-family applications. The earliest known test occurred in 2002 in

Ohio, where the homeowner combined waste vegetable oil with dry aggregate to create a low-cost and less polluting paving material for his 200-foot driveway. After five years, he reports the driveway is performing as well or better than petroleum-based materials.

Shell Oil Company paved two public roads in Norway in 2007 with vegetable-oil-based asphalt. Results of this study are still premature.

HALIK Asphalts LTD from Israel has been experimenting with recycled and secondary road building since 2003. The company is using various wastes such as vegetable fats & oils, wax and thermoplastic elastomers to build and repair roads. The results reported are so far satisfying.

On October 6, 2010, a bicycle path in Des Moines, Iowa, was paved with bio-oil based asphalt through a partnership between Iowa State University, the City of Des Moines, and Avello Bioenergy Inc. Research is being conducted on the asphalt mixture, derived from plants and trees to replace petroleum-based mixes. Bioasphalt is a registered trademark of Avello Bioenergy Inc.

Dr. Elham H. Fini, at North Carolina A&T University, has been spearheading research that has successfully produced bio asphalt from swine manure.

References

- Sixta, Herbert, ed. (2006). Handbook of pulp. 1. Winheim, Germany: Wiley-VCH. pp. 79–88. ISBN 3-527-30997-7.
- Hunter, Louis C.; Bryant, Lynwood (1991). A History of Industrial Power in the United States, 1730-1930, Vol. 3: The Transmission of Power. Cambridge, Massachusetts, London: MIT Press. ISBN 0-262-08198-9.
- "Biogas Flows Through Germany's Grid Big Time - Renewable Energy News Article". 14 March 2012. Archived from the original on 14 March 2012. Retrieved 17 June 2016.
- "404 - Seite nicht gefunden auf Server der Fachagentur Nachwachsende Rohstoffe e.V.: FNR" (PDF). Retrieved 17 June 2016.
- "World's large river deltas continue to degrade from human activity". News Center. Retrieved 2016-02-24.
- "National Grid: Half Britain's homes could be heated by renewable gas, says National Grid". Retrieved 15 May 2015.
- National Film Board of Canada. "Bate's Car: Sweet as a Nut". NFB.ca. Retrieved 15 May 2015.
- "GPS Renewables - Waste management through biogas - GPS Renewables". GPS Renewables. Retrieved 15 May 2015.
- "Biogas plants provide cooking and fertiliser". Ashden Awards, sustainable and renewable energy in the UK and developing world. Retrieved 15 May 2015.
- "ICIS Innovation Award Winners, Renmatix and Virent, Announce Collaboration On Bio-based Packaging", Retrieved 23 June 2015.
- "McKinsey & Company Releases New Data About Growth in Industrial Biotech Sector at World

Congress" Retrieved 23 June 2015.

- Third Annual World Congress on Industrial Biotechnology and Bioprocessing, Toronto, ON, July 11–14, 2006 Retrieved 23 June 2015.

- "McKinsey & Company Releases New Data About Growth in Industrial Biotech Sector at World Congress" Retrieved 30 July 2015.

- "Transforming Greenhouse Gas Emissions into Energy" (PDF). WIPO Green Case Studies, 2014. World Intellectual Property Organization. 2014. Retrieved 6 April 2015.

- "Launch of New 'Ene-Farm' Home Fuel Cell Product More Affordable and Easier to Install - Headquarters News - Panasonic Newsroom Global". Retrieved 15 May 2015.

- Elsevier Ltd, The Boulevard, Langford Lane, Kidlington, Oxford, OX5 1GB, United Kingdom. "Micro CHP report powers heated discussion about UK energy future". Retrieved 15 May 2015.

- National Non-Food Crops Centre. "NNFCC Renewable Fuels and Energy Factsheet: Anaerobic Digestion", Retrieved on 2011-02-16.

- "Cold climates no bar to biogas production". New Scientist. London: Sunita Harrington. 6 November 2010. p. 14. Retrieved 4 February 2011.

Utilization of Plants in Biomass

Plants in a number of ways are utilized in biomass. Some of the applications are lignocellulosic biomass, energy crop and panicum virgatum. Plant dry matter is referred to as lignocellulosic biomass whereas plants with little maintenance are used as biofuels are called energy crops. This section will provide an integrated understanding on the utilization of plants in biomass production.

Lignocellulosic Biomass

Lignocellulose refers to plant dry matter (biomass), so called lignocellulosic biomass. It is the most abundantly available raw material on the Earth for the production of biofuels, mainly bio-ethanol. It is composed of carbohydrate polymers (cellulose, hemicellulose), and an aromatic polymer (lignin). These carbohydrate polymers contain different sugar monomers (six and five carbon sugars) and they are tightly bound to lignin. Lignocellulosic biomass can be broadly classified into virgin biomass, waste biomass and energy crops. Virgin biomass includes all naturally occurring terrestrial plants such as trees, bushes and grass. Waste biomass is produced as a low value byproduct of various industrial sectors such as agriculture (corn stover, sugarcane bagasse, straw etc.) and forestry (saw mill and paper mill discards). Energy crops are crops with high yield of lignocellulosic biomass produced to serve as a raw material for production of second generation biofuel; examples include switch grass (Panicum virgatum) and Elephant grass.

Dedicated Energy Crops

Many crops are of interest for their ability to provide high yields of biomass and can be harvested multiple times each year. These include poplar trees and Miscanthus giganteus. The premier energy crop is sugarcane, which is a source of the readily fermentable sucrose and the lignocellulosic by-product bagasse.

Application

Pulp and Paper Industry

Lignocellulosic biomass is the feedstock for the pulp and paper industry. This energy-intensive industry focuses on the separation of the lignin and cellulosic fractions of the biomass.

Biofuels

Lignocellulosic biomass, in the form of wood fuel, has a long history as a source of energy. Since the middle of the 20th century, the interest of biomass as a precursor to *liquid* fuels has increased. To be specific, the fermentation of lignocellulosic biomass to ethanol is an attractive route to fuels that supplements the fossil fuels. Biomass is a carbon-neutral source of energy: Since it comes from plants, the combustion of ligno-cellulosic ethanol produces no net carbon dioxide into the earth's atmosphere. Aside from ethanol, many other lignocellulose-derived fuels are of potential interest, including butanol, dimethylfuran, and gamma-Valerolactone.

One barrier to the production of ethanol from biomass is that the sugars necessary for fermentation are trapped inside the lignocellulose. Lignocellulose has evolved to re-sist degradation and to confer hydrolytic stability and structural robustness to the cell walls of the plants. This robustness or "recalcitrance" is attributable to the crosslinking between the polysaccharides (cellulose and hemicellulose) and the lignin via ester and ether linkages. Ester linkages arise between oxidized sugars, the uronic acids, and the phenols and phenylpropanols functionalities of the lignin. To extract the fermentable sugars, one must first disconnect the celluloses from the lignin, and then use acid or enzymatic methods to hydrolyze the newly freed celluloses to break them down into simple monosaccharides. Another challenge to biomass fermentation is the high per-centage of pentoses in the hemicellulose, such as xylose, or wood sugar. Unlike hexoses such as glucose, pentoses are difficult to ferment. The problems presented by the lignin and hemicellulose fractions are the foci of much contemporary research.

A large sector of research into the exploitation of lignocellulosic biomass as a feedstock for bio-ethanol focuses particularly on the fungus Trichoderma reesei, known for its cellulolytic abilities. Multiple avenues are being explored including the design of an optimised cocktail of cellulases and hemicellulases isolated from *T. reesei*, as well as genetic-engineering-based strain improvement to allow the fungus to simply be placed in the presence of lignocellulosic biomass and break down the matter into D-glucose monomers. Strain improvement methods have led to strains capable of producing sig-nificantly more cellulases than the original QM6a isolate; certain industrial strains are known to produce up to 100g of cellulase per litre of fungus thus allowing for maximal extraction of sugars from lignocellulosic biomass. These sugars can then be fermented, leading to bio-ethanol.

Panicum Virgatum

Panicum virgatum, commonly known as switchgrass, is a perennial warm season bunchgrass native to North America, where it occurs naturally from 55°N latitude in Canada southwards into the United States and Mexico. Switchgrass is one of the dominant species of the central North American tallgrass prairie and can be found in remnant prairies, in native grass pastures, and naturalized along roadsides. It is used primarily for soil conservation, forage production, game cover, as an ornamental grass, and more recently as a biomass crop for ethanol and butanol, in phytoremediation projects, fiber, electricity, and heat production and for biosequestration of atmospheric carbon dioxide.

Other common names for switchgrass include tall panic grass, Wobsqua grass, black-bent, tall prairiegrass, wild redtop, thatchgrass, and Virginia switchgrass.

Description

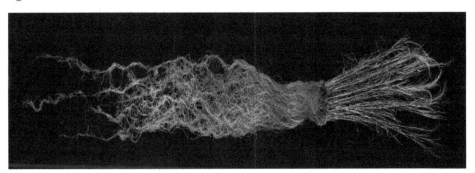

Root system of switchgrass grown at the Land Institute

Switchgrass is a hardy, deep-rooted, perennial rhizomatous grass that begins growth in late spring. It can grow up to 2.7 m high, but is typically shorter than big bluestem grass or indiangrass. The leaves are 30–90 cm long, with a prominent midrib. Switchgrass uses C_4 carbon fixation, giving it an advantage in conditions of drought and high temperature. Its flowers have a well-developed panicle, often up to 60 cm long, and it bears a good crop of seeds. The seeds are 3–6 mm long and up to 1.5 mm wide, and are developed from a single-flowered spikelet. Both glumes are present and well developed. When ripe, the seeds sometimes take on a pink or dull-purple tinge, and turn golden brown with the foliage of the plant in the fall. Switchgrass is both a perennial and self-seeding crop, which means farmers do not have to plant and reseed after annual harvesting. Once established, a switchgrass stand can survive for ten years or longer. Unlike corn, switchgrass can grow on marginal lands and requires relatively modest levels of chemical fertilizers. Overall, it is considered a resource-efficient, low-input crop for producing bioenergy from farmland.

Habitat

Much of North America, especially the prairies of the Midwestern United States, was once prime habitat to vast swaths of native grasses, including switchgrass, indiangrass (*Sorghastrum nutans*), eastern gamagrass (Tripsacum dactyloides), big bluestem (Andropogon gerardi), little bluestem (Schizachyrium scoparium), bluestem prairie (Andropogon panicum sorghastrum) and others. As European settlers began spreading west across the continent, the native grasses were plowed under and the land converted to crops such as corn, wheat, and oats. Introduced grasses such as fescue, bluegrass, and orchardgrass also replaced the native grasses for use as hay and pasture for cattle.

Distribution

Switchgrass is a versatile and adaptable plant. It can grow and even thrive in many weather conditions, lengths of growing seasons, soil types, and land conditions. Its distribution spans south of latitude 55°N from Saskatchewan to Nova Scotia, south over most of the United States east of the Rocky Mountains, and further south into Mexico. As a warm-season perennial grass, most of its growth occurs from late spring through early fall; it becomes dormant and unproductive during colder months. Thus, the productive season in its northern habitat can be as short as three months, but in the southern reaches of its habitat the growing season may be as long as eight months, around the Gulf Coast area.

Switchgrass is a diverse species, with striking differences between plants. This diversity, which presumably reflects evolution and adaptation to new environments as the species spread across the continent, provides a range of valuable traits for breeding programs. Switchgrass has two distinct forms, or "cytotypes": the lowland cultivars, which tend to produce more biomass, and the upland cultivars, which are generally of more northern origin, more cold-tolerant, and therefore usually preferred in northern areas. Upland switchgrass types are generally shorter ($\leq$ 2.4 m, tall) and less coarse than lowland types. Lowland cultivars may grow to $\geq$ 2.7 m, in favorable environments. Both upland and lowland cultivars are deeply rooted (> 1.8 m, in favorable soils) and have short rhizomes. The upland types tend to have more vigorous rhizomes, so the lowland cultivars may appear to have a bunchgrass habit, while the upland types tend to be more sod-forming. Lowland cultivars appear more plastic in their morphology, produce larger plants if stands become thin or when planted in wide rows, and they seem to be more sensitive to moisture stress than upland cultivars.

In native prairies, switchgrass is historically found in association with several other important native tallgrass prairie plants, such as big bluestem, indiangrass, little bluestem, sideoats grama, eastern gamagrass, and various forbs (sunflowers, gayfeather, prairie clover, and prairie coneflower). These widely adapted tallgrass species once occupied millions of hectares.

Switchgrass' suitability for cultivation in the Gran Chaco is being studied by Argentina's Instituto Nacional de Tecnología Agropecuaria (INTA).

Establishment and Management

Switchgrass can be grown on land considered unsuitable for row crop production, including land that is too erodible for corn production, as well as sandy and gravelly soils in humid regions that typically produce low yields of other farm crops. No single method of establishing switchgrass can be suggested for all situations. The crop can be established both by no-till and conventional tillage. When seeded as part of a diverse mixture, planting guidelines for warm-season grass mixtures for conservation plantings should be followed. Regional guidelines for growing and managing switchgrass for bioenergy or conservation plantings are available. Several key factors can increase the likelihood of success for establishing switchgrass. These include:

- Planting switchgrass after the soil is well warmed during the spring.

- Using seeds that are highly germinable and planting 0.6 - 1.2 cm deep, or up to 2 cm deep in sandy soils.

- Packing or firming the soil both before and after seeding.

- Providing no fertilization at planting to minimize competition.

- Controlling weeds with chemical and/or cultural control methods.

Mowing and properly labeled herbicides are recommended for weed control. Chemical weed control can be used in the fall prior to establishment, or before or after planting. Weeds should be mowed just above the height of the growing switchgrass. Hormone herbicides, such as 2,4-D, should be avoided as they are known to reduce development of switchgrass when applied early in the establishing year. Plantings that appear to have failed due to weed infestations are often wrongly assessed, as the failure is often more apparent than real. Switchgrass stands that are initially weedy commonly become well established with appropriate management in subsequent years. Once established, switchgrass can take up to three years to reach its full production potential. Depending on the region, it can typically produce 1/4 to 1/3 of its yield potential in its first year and 2/3 of its potential in the year after seeding.

After establishment, switchgrass management will depend on the goal of the seeding. Historically, most switchgrass seedings have been managed for the Conservation Reserve Program in the US. Disturbance such as periodic mowing, burning, or disking is required to optimize the stand's utility for encouraging biodiversity. Increased attention is being placed on switchgrass management as an energy crop. Generally, the crop requires modest application of nitrogen fertilizer, as it is not a heavy feeder. Typical nitrogen (N) content of senescent material in the fall is 0.5% N. Fertilizer nitrogen applications of about 5 kg N/hectare (ha) applied for each tonne of biomass removed is a general guideline. More specific recommendations for fertilization are available regionally in North America. Herbicides are not often used on switchgrass after the seeding year, as the crop is generally quite competitive with weeds. Most bioenergy conversion processes for switchgrass, including those for cellulosic ethanol and pellet fuel production, can generally accept some alternative species in the harvested biomass. Stands of switchgrass should be harvested no more than twice per year, and one cutting often provides as much biomass as two. Switchgrass can be harvested with the same field equipment used for hay production, and it is well-suited to baling or bulk field harvesting. If its biology is properly taken into consideration, switchgrass can offer great potential as an energy crop.

Uses

Switchgrass can be used as a feedstock for biomass energy production, as ground cover for soil conservation, and to control erosion, for forages and grazing, as game cover, and as feedstock for biodegradable plastics. It can be used by cattle farmers for hay and pasture and as a substitute for wheat straw in many applications, including livestock bedding, straw bale housing, and as a substrate for growing mushrooms.

Panicum virgatum 'Heavy Metal', an ornamental switchgrass, in early summer

Additionally, switchgrass is grown as a drought-resistant ornamental grass in average to wet soils and in full sun to part shade.

Bioenergy

Switchgrass has been researched as a renewable bioenergy crop since the mid-1980s, because it is a native perennial warm season grass with the ability to produce moderate to high yields on marginal farmlands. It is now being considered for use in several bioenergy conversion processes, including cellulosic ethanol production, biogas, and direct combustion for thermal energy applications. The main agronomic advantages of switchgrass as a bioenergy crop are its stand longevity, drought and flooding tolerance, relatively low herbicide and fertilizer input requirements, ease of management, hardiness in poor soil and climate conditions, and widespread adaptability in temperate climates. In some warm humid southern zones, such as Alabama, it has the ability to produce up to 25 oven-dry tonnes per hectare (ODT/ha). A summary of switchgrass yields across 13 research trial sites in the United States found the top two cultivars in each trial to yield 9.4 to 22.9 t/ha, with an average yield of 14.6 ODT/ha. However, these yields were recorded on small plot trials, and commercial field sites could be expected to be at least 20% lower than these results. In the United States, switchgrass yields appear to be highest in warm humid regions with long growing seasons such as the US Southeast and lowest in the dry short season areas of the Northern Great Plains. The energy inputs required to grow switchgrass are favorable when compared with annual seed bearing crops such as corn, soybean, or canola, which can require relatively high energy inputs for field operations, crop drying, and fertilization. Whole plant herbaceous perennial C4 grass feedstocks are desirable biomass energy feedstocks, as they require fewer fossil energy inputs to grow and effectively capture solar energy because of their C4 photosynthetic system and perennial nature. One study cites it takes from 0.97 to 1.34 GJ to produce 1 tonne of switchgrass, compared with 1.99 to 2.66 GJ to produce 1 tonne of corn. Another study found that switchgrass uses 0.8 GJ/ODT of fossil energy compared to grain corn's 2.9 GJ/ODT. Given that switchgrass contains approximately 18.8 GJ/ODT of biomass, the energy output-to-input ratio for the crop can be up to 20:1. This highly favorable ratio is attributable to its relatively high energy output per hectare and low energy inputs for production.

Considerable effort is being expended in developing switchgrass as a cellulosic ethanol crop in the USA. In George W. Bush's 2006 State of the Union Address, he proposed using switchgrass for ethanol; since then, over US$100 million has been invested into researching switchgrass as a potential biofuel source. Switchgrass has the potential to produce up to 380 liters of ethanol per tonne harvested. However, current technology for herbaceous biomass conversion to ethanol is about 340 liters per tonne. In contrast, corn ethanol yields about 400 liters per tonne. There are substantial efforts being made to increase the yield of corn derived ethanol:

(Corn-) Ethanol yield has already improved from 2.4 gallons per bushel in the 1980s to 2.8 gallons in modern plants. Corn hybrids developed specifically for ethanol production have demonstrated ethanol yield increases of 2.7 percent—and using the cellulose (fiber) in the corn kernel, in addition to the starch, could increase yield by another 10 to 13 percent. With this combination of hybrid and process optimization, theoretical yields of 3.51 gallons of ethanol per bushel are within reason—with no negative impact on protein or oil content for animal feed uses of the distillers grains.

Process improvements for the conventional corn-based industry are based on new technologies such as cavitation, new catalysis and new enzymes.

The main advantage of using switchgrass over corn as an ethanol feedstock is its cost of production is generally about 1/2 that of grain corn, and more biomass energy per hectare can be captured in the field. Thus, switchgrass cellulosic ethanol should give a higher yield of ethanol per hectare at lower cost. However, this will depend on whether the cost of constructing and operating cellulosic ethanol plants can be reduced considerably. The switchgrass ethanol industry energy balance is also considered to be substantially better than that of corn ethanol. During the bioconversion process, the lignin fraction of switchgrass can be burned to provide sufficient steam and electricity to operate the biorefinery. Studies have found that for every unit of energy input needed to create a biofuel from switchgrass, four units of energy are yielded. In contrast, corn ethanol yields about 1.28 units of energy per unit of energy input. A recent study from the Great Plains indicated that for ethanol production from switchgrass, this figure is 6.4, or alternatively, that 540% more energy was contained in the ethanol produced than was used in growing the switchgrass and converting it to liquid fuel. However, there remain commercialization barriers to the development of cellulosic ethanol technology. Projections in the early 1990s for commercialization of cellulosic ethanol by the year 2000 have not been met. The commercialization of cellulosic ethanol is thus proving to be a significant challenge, despite noteworthy research efforts.

Thermal energy applications for switchgrass appear to be closer to near-term scale-up than cellulosic ethanol for industrial or small-scale applications. For example, switchgrass can be pressed into fuel pellets that are subsequently burned in pellet stoves used to heat homes (which typically burn corn or wood pellets). Switchgrass has been widely tested as a substitute for coal in power generation. The most widely studied project to date has been the Chariton Valley Project in Iowa. The Show-Me-Energy Cooperative (SMEC) in Missouri is using switchgrass and other warm-season grasses, along with wood residues, as feedstocks for pellets used for the firing of a coal-fired power plant. In Eastern Canada, switchgrass is being used on a pilot scale as a feedstock for commercial heating applications. Combustion studies have been undertaken and it appears to be well-suited as a commercial boiler fuel. Research is also being undertaken to develop switchgrass as a pellet fuel because of lack of surplus wood residues in eastern Canada, as a slowdown in the forest products industry in 2009 is now resulting in wood pellet shortages throughout Eastern North America. Generally speaking, the direct firing of switchgrass for thermal applications can provide the highest net energy gain and energy output-to-input ratio of all switchgrass bioconversion processes. Research has found switchgrass, when pelletized and used as a solid biofuel, is a good candidate for displacing fossil fuels. Switchgrass pellets were identified to have a 14.6:1 energy output-to-input ratio, which is substantially better than that for liquid biofuel options from farmland. As a greenhouse gas mitigation strategy, switchgrass pellets were found to be an effective means to use farmland to mitigate greenhouse gases on the order of 7.6-13 tonnes per hectare of CO_2. In contrast, switchgrass cellulosic ethanol and corn ethanol were found to mitigate 5.2 and 1.5 tonnes of CO_2 per hectare, respectively.

Historically, the major constraint to the development of grasses for thermal energy applications has been the difficulty associated with burning grasses in conventional boilers, as biomass quality problems can be of particular concern in combustion applications. These technical problems now appear to have been largely resolved through crop management practices such as fall mowing and spring harvesting that allow for leaching to occur, which leads to fewer aerosol-forming compounds (such as K and Cl) and N in the grass. This reduces clinker formation and corrosion, and enables switchgrass to be a clean combustion fuel source for use in smaller combustion appliances. Fall harvested grasses likely have more application for larger commercial and industrial boilers. Switchgrass is also being used to heat small industrial and farm buildings in Germany and China through a process used to make a low quality natural gas substitute.

Bai et al. (2010) conducted a study to analyze the environmental sustainability of using switchgrass plant material as a feedstock for ethanol production. Life cycle analysis was used to make this assessment. They compared efficiency of E10, E85, and ethanol with gasoline. They took into account air and water emissions associated with growing, managing, processing and storing the switchgrass crop. They also factored in the transportation of the stored switchgrass to the ethanol plant where they assumed the distance was 20 km. The reductions in global warming potential by using E10 and E85 were 5 and 65%, respectively. Their models also suggested that the "human toxicity potential" and "eco-toxicity potential" were substantially greater for the high ethanol fuels (i.e., E85 and ethanol) than for gasoline and E10.

In 2014, a genetically altered form of the bacterium Caldicellulosiruptor bescii was created which can cheaply and efficiently turn switchgrass into ethanol.

Biodegradable Plastics Production

In a novel application, US scientists have genetically modified switchgrass to enable it to produce polyhydroxybutyrate, which accumulates in beadlike granules within the plant's cells. In preliminary tests, the dry weight of a plants leaves were shown to comprise up to 3.7% of the polymer. Such low accumulation rates do not, as of 2009, allow for commercial use of switchgrass as a biosource.

Soil Conservation

Switchgrass is useful for soil conservation and amendment, particularly in the United States and Canada, where switchgrass is endemic. Switchgrass has a deep fibrous root system – nearly as deep as the plant is tall. Since it, along with other native grasses and forbs, once covered the plains of the United States that are now the Corn Belt, the effects of the past switchgrass habitat have been beneficial, lending to the fertile farmland that exists today. The deep fibrous root systems of switchgrass left a deep rich layer of organic matter in the soils of the Midwest, making those mollisol soils some of the most productive in the world. By returning switchgrass and other perennial prairie grasses as an agricultural crop, many marginal soils may benefit from increased levels of organic material, permeability, and fertility, due to the grass's deep root system.

Soil erosion, both from wind and water, is of great concern in regions where switchgrass grows. Due to its height, switchgrass can form an effective wind erosion barrier. Its root system, also, is excellent for holding soil in place, which helps prevent erosion from flooding and runoff. Some highway departments (for example, KDOT) have used switchgrass in their seed mixes when re-establishing growth along roadways. It can also be used on strip mine sites, dikes, and pond dams. Conservation districts in many parts of the United States use it to control erosion in grass waterways because of its ability to anchor soils while providing habitat for wildlife.

Forages and Grazing

Switchgrass is an excellent forage for cattle; however, it has shown toxicity in horses, sheep, and goats through chemical compounds known as saponins, which cause photosensitivity and liver damage in these animals. Researchers are continuing to learn more about the specific conditions under which switchgrass causes harm to these species, but until more is discovered, it is recommended switchgrass not be fed to them. For cattle, however, it can be fed as hay, or grazed.

Grazing switchgrass calls for watchful management practices to ensure survival of the stand. It is recommended that grazing begin when the plants are about 50 cm tall, and that grazing be discontinued when the plants have been eaten down to about 25 cm, and to rest the pasture 30 – 45 days between grazing periods. Switchgrass becomes stemmy and unpalatable as it matures, but during the target grazing period, it is a favorable forage with a relative feed value (RFV) of 90-104. The grass's upright growth pattern places its growing point off the soil surface onto its stem, so leaving 25 cm of stubble is important for regrowth. When harvesting switchgrass for hay, the first cutting occurs at the late boot stage – around mid-June. This should allow for a second cutting in mid-August, leaving enough regrowth to survive the winter.

Game Cover

Switchgrass is well-known among wildlife conservationists as good forage and habitat for upland game bird species, such as pheasant, quail, grouse, and wild turkey, and song birds, with its plentiful small seeds and tall cover. A study published in 2015 has shown that switchgrass, when grown in a traditional monoculture, has an adverse impact on some wildlife. Depending on how thickly switchgrass is planted, and what it is partnered with, it also offers excellent forage and cover for other wildlife across the country. For those producers who have switchgrass stands on their farm, it is considered an environmental and aesthetic benefit due to the abundance of wildlife attracted by the switchgrass stands. Some members of Prairie Lands Bio-Products, Inc. in Iowa have even turned this benefit into a profitable business by leasing their switchgrass land for hunting during the proper seasons. The benefits to wildlife can be extended even in large-scale agriculture through the process of strip harvesting, as recommended by The Wildlife Society, which suggests that rather than harvesting an entire field at once, strip harvesting could be practiced so that the entire habitat is not removed, thereby protecting the wildlife inhabiting the switchgrass.

Energy Crop

A Department for Environment, Food and Rural Affairs energy crops scheme plantation in the United Kingdom. Energy crops of this sort can be used in conventional power stations or specialised electricity generation units, reducing the amount of fossil fuel-derived carbon dioxide emissions.

An energy crop is a plant grown as a low-cost and low-maintenance harvest used to make biofuels, such as bioethanol, or combusted for its energy content to generate electricity or heat. Energy crops are generally categorized as woody or herbaceous plants; many of the latter are grasses (Graminaceae).

Commercial energy crops are typically densely planted, high-yielding crop species, processed to bio-fuel and burnt to generate power. Woody crops such as willow or poplar are widely utilised, as well as temperate grasses such as Miscanthus and Pennisetum purpureum (both known as elephant grass). If carbohydrate content is desired for the production of biogas, whole-crops such as maize, Sudan grass, millet, white sweet clo-ver and many others, can be made into silage and then converted into biogas.

Through genetic modification and application of biotechnology plants can be manipu-lated to create greater yields, reduce associated costs and require less water. However, high energy yield can be realized with existing cultivars.

Types of Energy Crops

By State

Solid Biomass

Energy generated by burning plants grown for the purpose, often after the dry matter is pelletized. Energy crops are used for firing power plants, either alone or co-fired with other fuels. Alternatively they may be used for heat or combined heat and power (CHP) production.

Elephant grass (Miscanthus sinensis) is an experimental energy crop

(Note: The terms biofuel, biomass, and so on, are often used interchangeably.)

To cover the increasing requirements of woody biomass, short rotation coppice (SRC) were applied to agricultural sites. Within this crpoping systems fast growing tree species like willows and poplars are planted in growing cycles of three to five years. The cultivation of this cultures is dependent on wet soil conditions and could be an alternative for moist field sieds. However, a influence on local water conditions could not be excluded. This indicates that an establishment should exclude the vicinity to vulnarable wetland ecosystems.

Gas Biomass (Methane)

Anaerobic digesters or biogas plants can be directly supplemented with energy crops once they have been ensiled into silage. The fastest growing sector of German biofarm-ing has been in the area of "Renewable Energy Crops" on nearly 500,000 ha of land (2006). Energy crops can also be grown to boost gas yields where feedstocks have a low energy content, such as manures and spoiled grain. It is estimated that the energy yield presently of bioenergy crops converted via silage to methane is about 2 GWh/km^2. Small mixed cropping enterprises with animals can use a portion of their acreage to grow and convert energy crops and sustain the entire farms energy requirements with about 1/5 the acreage. In Europe and especially Germany, however, this rapid growth has occurred only with substantial government support, as in the German bonus system for renewable energy. Similar developments of integrating crop farming and bioenergy production via silage-methane have been almost entirely overlooked in N. America, where political and structural issues and a huge continued push to centralize energy production has overshadowed positive developments.

Liquid Biomass

Biodiesel

European production of biodiesel from energy crops has grown steadily in the last decade, principally focused on rapeseed used for oil and energy. Production of oil/biodiesel from rape covers more than 12,000 km^2 in Germany alone, and has doubled in the past 15 years. Typical yield of oil as pure biodiesel may be is 100,000 L/km^2 or more, making biodiesel crops economically attractive, provided sustainable crop rotations exist that are nutrient-balanced and preventative of the spread of disease such as clubroot. Biodiesel yield of soybeans is significantly lower than that of rape.

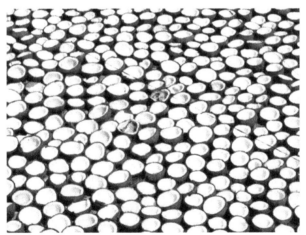

Coconuts sun-dried in Kozhikode, Kerala for making copra, the dried meat, or kernel, of the coconut. Coconut oil extracted from it has made copra an important agricultural commodity for many coconut-producing countries. It also yields coconut cake which is mainly used as feed for livestock.

Typical oil extractable by weight	
Crop	**Oil %**
copra	62
castor seed	50
sesame	50

ation with cellulosic bioethanol in America as the agricultural structure supporting biomethane is absent in many regions, with no credits or bonus system in place. Consequently, a lot of private money and investor hopes are being pinned on marketable and patentable innovations in enzyme hydrolysis and the like.

Bioethanol also refers to the technology of using principally corn (maize seed) to make ethanol directly through fermentation, a process that under certain field and process conditions can consume as much energy as is the energy value of the ethanol it produces, therefore being non-sustainable. New developments in converting grain stillage (referred to as distillers grain stillage or DGS) into biogas energy looks promising as a means to improve the poor energy ratio of this type of bioethanol process.

By Dedication

Dedicated energy crops are non-food energy crops as giant miscanthus, switchgrass, jatropha, fungi, and algae. Dedicated energy crops are promising cellulose sources that can be sustainably produced in many regions of the United States.

Additionally, Byproducts (green waste) of food and non-food energy crops can be used to produce various biofuels.

References

- Ball, D.M.; Hoveland, C.S.; Lacefield, G.D. (2002). Southern Forages (3rd ed.). International Plant Nutrition Institute. p. 26. ISBN 978-0-9629598-3-7.

- Ara Kirakosyan; Peter B. Kaufman (2009-08-15). Recent Advances in Plant Biotechnology. p. 169. ISBN 9781441901934. Retrieved 14 February 2013.

- "Direct conversion of plant biomass to ethanol by engineered Caldicellulosiruptor bescii". Pnas. org. Retrieved 2014-06-04.

- Index of Species Information (2013). "Species: Panicum virgatum". US Forest Service. Archived from the original on 2012-11-08. Retrieved 2013-12-13.

Permissions

We would like to thank the editorial team for lending their expertise to make the book truly unique. They have played a crucial role in the development of this book. Without their invaluable contributions this book wouldn't have been possible. They have made vital efforts to compile up to date information on the varied aspects of this subject to make this book a valuable addition to the collection of many professionals and students.

This book was conceptualized with the vision of imparting up-to-date and integrated information in this field. To ensure the same, a matchless editorial board was set up. Every individual on the board went through rigorous rounds of assessment to prove their worth. After which they invested a large part of their time researching and compiling the most relevant data for our readers.

The editorial board has been involved in producing this book since its inception. They have spent rigorous hours researching and exploring the diverse topics which have resulted in the successful publishing of this book. They have passed on their knowledge of decades through this book. To expedite this challenging task, the publisher supported the team at every step. A small team of assistant editors was also appointed to further simplify the editing procedure and attain best results for the readers.

Apart from the editorial board, the designing team has also invested a significant amount of their time in understanding the subject and creating the most relevant covers. They scrutinized every image to scout for the most suitable representation of the subject and create an appropriate cover for the book.

The publishing team has been an ardent support to the editorial, designing and production team. Their endless efforts to recruit the best for this project, has resulted in the accomplishment of this book. They are a veteran in the field of academics and their pool of knowledge is as vast as their experience in printing. Their expertise and guidance has proved useful at every step. Their uncompromising quality standards have made this book an exceptional effort. Their encouragement from time to time has been an inspiration for everyone.

The publisher and the editorial board hope that this book will prove to be a valuable piece of knowledge for students, practitioners and scholars across the globe.

Index

www.ingramcontent.com/pod-product-compliance
Lightning Source LLC
Jackson TN
JSHW052203130125
77033JS00004B/204